BATTLES OF THE NORTH COUNTRY

A VOLUME IN THE SERIES
Environmental History of the Northeast
EDITED BY
Anthony N. Penna
Richard W. Judd

BATTLES OF THE NORTH COUNTRY

Wilderness Politics and
Recreational Development
in the Adirondack State Park,
1920–1980

JONATHAN D. ANZALONE

University of Massachusetts Press
Amherst and Boston

Copyright © 2018 by University of Massachusetts Press
All rights reserved
Printed in the United States of America

ISBN 978-1-62534-364-2 (paper); 363-5 (hardcover)

Designed by Jennifer Jackowitz
Set in Adobe Garamond Pro

Cover design by Patricia Duque Campos
Cover photo: New York State Archives, Division of Visual Instruction, Instructional Lantern Slides, A3045, no. A10340.

Library of Congress Cataloging-in-Publication Data

Names: Anzalone, Jonathan D., author.
Title: Battles of the north country : wilderness politics and recreational development in the Adirondack State Park, 1920–1980 / Jonathan D. Anzalone.
Other titles: Wilderness politics and recreational development in the Adirondack State Park, 1920–1980
Description: Amherst, MA : University of Massachusetts Press, [2018] | Series: Environmental history of the Northeast | Includes bibliographical references and index. |
Identifiers: LCCN 2018019134 (print) | LCCN 2018026673 (ebook) | ISBN 9781613766088 (e-book) | ISBN 9781613766095 (e-book) | ISBN 9781625343642 (pbk.) | ISBN 9781625343635 (hardcover)
Subjects: LCSH: Adirondack Park (N.Y.)—History. | Outdoor recreation—New York (State)—Adirondack Park—History. | Wilderness areas—New York (State)—Adirondack Mountains—History. | Land use—New York (State)—Adirondack Park—Planning. | Land use—New York (State)—Adirondack Park—Citizen participation. | Conservation of natural resources—New York (State)—Adirondack Park—History. | New York (State)—Politics and government—20th century.
Classification: LCC F127.A2 (ebook) | LCC F127.A2 A278 2018 (print) | DDC 974.7/5—dc23
LC record available at https://lccn.loc.gov/2018019134

British Library Cataloguing-in-Publication Data
A catalog record for this book is available from the British Library.

FOR SALLY AND BENJAMIN

CONTENTS

Acknowledgments ix

Introduction
The Adirondack Park as a Modern Wilderness Playground 1

Chapter 1. Olympic Transformations, Part I
The Re-creation of Recreation and the 1932 Winter Games in Lake Placid 11

Chapter 2. Cities of Tents
Development of Adirondack Campgrounds during the Interwar Years 36

Chapter 3. A Mountain to Climb
The Transformation of Whiteface Mountain and the Future of the Adirondacks, 1925–1945 61

Chapter 4. A Mountain for All Seasons?
New York State and Skiing on Whiteface Mountain, 1945–1971 82

Chapter 5. Adirondack Sprawl
From the Northway to the Creation of the Adirondack Park Agency, 1959–1972 105

Chapter 6. "There Was Once an Adirondack Park"
The Struggle over the Exurbanization of the Adirondack Park, 1971–1980   135

Chapter 7. Olympic Transformations, Part II
The 1980 Winter Games in Lake Placid   171

Conclusion   203

Notes   209
Index   267

ACKNOWLEDGMENTS

I have incurred many debts while working on this book, especially to the mentors who helped me improve as a historian and a writer. Christopher Sellers constantly pushed me to be bold with my claims, and my work is stronger as a result. Nancy Tomes and Shirley Lim helped me sharpen my ideas and find new ways to connect my work with other historical fields. I also give thanks to April Masten, Donna Rilling, Thomas Klubock, Kathleen Wilson, Wilbur Miller, and Ann Norton Greene for their help and support.

I would be remiss if I did not thank the archivists and librarians who helped me track down sources and sift through the archives. Jerry Pepper and Ivy Gocker were pleasant and knowledgeable guides through the collections of the Adirondack Museum Library. Jim Folts and his staff at the New York State Archives provided valuable assistance as well. I am also grateful to the New York State Archives Partnership Trust, which awarded me the Larry J. Hackman Research Residency to help defray the costs of my research trips to Albany. I owe thanks to staff at the Department of Environmental Conservation, the

Adirondack Park Agency, the New York Public Library, and the Stony Brook University Library.

Many thanks to the staff at the University of Massachusetts Press, especially Brian Halley and Tony Penna. I am grateful as well to the anonymous readers of the press. Gerry Krieg of Krieg Mapping prepared the maps in this book, and I am thankful for his fine work and patience.

Among my colleagues, I must single out Ron Van Cleef and Jeff Hall, whose excellence as historians is exceeded by their generosity as friends. Jeff, in particular, was generous in collecting and sharing sources he acquired during his trips to the Adirondacks. Jeff's perspective as a North Country native was also indispensable, and he was always willing to listen to my ideas, read my drafts, and offer his input.

Even with the assistance of the individuals and institutions thanked above, words no doubt fail me throughout this book; and they could not possibly convey my gratitude toward my family. My siblings, Nick, Marc, Greg, Stacey, and Ashleigh, are wonderful people whose example I attempt to emulate every day. Through their words and example, my parents, Donna and Anthony, have taught me to work hard and to cultivate my intellect and imagination. This book was long in the making, and unfortunately my mother passed away before its publication. Still, my achievements, however meager, would not be possible without her unconditional love and support, and I continue to strive every day to make her proud.

I save my final thank-you for my wife, Sally. Though she joined me late in the process of writing this book, her love, patience, and encouragement helped push me across the finish line. I could not ask for a more caring and supportive person with whom to share my life. Thank you, Sally, for everything—especially for our wonderful son, Benjamin.

BATTLES OF THE
NORTH COUNTRY

INTRODUCTION

The Adirondack Park as a Modern Wilderness Playground

"Without something happening, we're going to be a ghost town." So said Roger Amell, the town supervisor of Tupper Lake, in February 2012. Amell made this comment soon after something had in fact happened. The Adirondack Park Agency (APA), the state body responsible for managing both private and state lands in the park, gave its stamp of approval to the six thousand-acre Adirondack Club and Resort, the largest project the agency had ever approved.[1]

Amell and other supporters of the proposed development would have to delay their celebrations, however. Environmental groups challenged the project, kicking off a two-year legal battle that concluded in December 2014 and ultimately cleared the way for the developers to begin work on the hotel, skiing area, golf course, and 650 luxury homes planned for the resort. Though the Adirondack Club and Resort's website promised that the project was "COMING SPRING 2016," the season passed without a single new home being built. Behind the delay were the developers' financial struggles. In spring 2016, they owed more than $250,000 to Franklin County in back taxes, still had not purchased all of the lands earmarked for development, and had not yet secured all of the necessary permits. While the search for new investors went on, Tupper Lake's full-time residents, holding on in a struggling rural community, were forced to wait for the hundreds of jobs the project was supposed to create.[2]

However the story of the Adirondack Club and Resort plays out, this ongoing narrative is a familiar one in the region—its first chapters written more than a century earlier. Tupper Lake may not be a ghost town, but the spirits of wilderness battles past continue to haunt the Adirondacks.

To understand the causes and consequences of the frequent battles over land use in the Adirondack Park, we must go back to the origins of state intervention in the region. Park-making in the Adirondacks, as the story of the Adirondack Club and Resort indicates, was not a nostalgic endeavor to preserve an unchanging wilderness, but rather a collaborative effort by state administrators, caretakers, private developers, residents, and recreation seekers to modernize and domesticate the Adirondack Park. Planners' notion of recreation as a set of wholesome activities pursued by urban and suburban families in an unthreatening environment significantly shaped their approach to park development. Though the Adirondacks may be best known and most frequently studied as a wilderness, it is imperative that we treat the park as a space containing a variety of landscapes at different stages on a spectrum from pristine to developed.[3] It is necessary to see the park as a home for people, flora, and fauna; a workplace for residents and state caretakers; and a vacation destination for a variety of recreation seekers. The park is best understood not as an inherently natural space, but rather as the hybrid product of hundreds of contested decision and actions.[4] In short, we must view the Adirondack Park as a *modern wilderness playground*.

The term modern wilderness playground captures the diversity and contested nature of the Adirondack Park. It is a unique place that contains private lands where 100,000 to 130,000 people have lived since its creation in 1892. About half of its acreage has been protected by the state since 1894, declared in the constitution as "forever wild." Describing it as a modern space enables the incorporation of amenity-rich recreational sites, neighborhoods with permanent residents, and vacation-home communities into the Adirondack story as common features of the landscape. The park has never been a static space, for its shape has been contingent upon the plans of politicians and state administrators, the designs of businesspeople, the whims of residents and recreation seekers, and the environment's ecological dynamism. Beginning in the 1920s, the addition of modern amenities signaled the creation of a new standard for park-making geared toward mass recreation and economic growth. As such, the state's substantial investment in tourism was not an attempt to erase permanent residents off the map; rather, it was an effort to stave off economic stagnation in the region, filling the void left by both absent and declining industries. In addition to these modern structures' material impact on the land, their effects on the region's economy, and their influence on vacationers' leisure-time habits, they tied New York's North Country ever closer to the city. These powerful links connecting urban areas to the countryside ultimately reinforced the latter's dependence on the former, and fostered a long and bitter debate over land-use policy.

Those urban-rural links failed to compensate for the sporadic development that forced Adirondackers to cater to recreation seekers for their economic survival. Even when, during the late nineteenth and early twentieth centuries, manufacturing and lumber companies and mining firms employed a number of people in the region, low pay and the seasonal nature of the work made eking out a living a significant challenge for so many Adirondack Park residents. The shuttering of some industrial firms during the early twentieth century and the accelerated deindustrialization that reshaped the economies of both New York's North Country and the United States at large after the Second World War only exacerbated locals' struggle to make ends meet. Outside the orbit of Adirondack industry, of course, residents of rural and wild areas had been living in industrial deserts long before the post–World War II urban crisis hollowed out cities from New York City to Binghamton to rust-belt Buffalo in their own state, and from Baltimore to Detroit and to Oakland nationwide.[5] Along the way, Adirondack residents learned and relearned the lesson that living in a modern wilderness playground came with a heavy burden.

Nonetheless, a focus on various stakeholders' efforts to make up for a lack of productive industry with ambitious recreational development does not mean we can ignore the "wilderness" that was the product of decisions made by landowners and the state going back to the nineteenth century. The wilderness was not fenced off and left alone, for it, too, was a playground: a space for leave-no-trace camping and hiking set apart from the more domesticated sections of the park. That wilderness, according to many observers, was situated on one end of an environmental spectrum, with the sprawling metropolis on the opposite end. What defined the Adirondack Park's history was not simply the presence of wilderness, but rather the park's mixed landscape featuring woodlands and communities that corresponded to almost every stage on the spectrum of environments. A close look at the history of the diverse landscapes and people in the region offers an opportunity for a wide-ranging study relevant to many different threads in the field of environmental history.

The work of environmental historian Karl Jacoby is integral to this study. In his *Crimes against Nature*, he examines Adirondack residents' "moral ecology," by which he means their folk traditions governing resource use that evolved in counterpoint to state conservation policies during the late nineteenth and early twentieth centuries.[6] Both Jacoby's work and this book combine social, political, and environmental history to analyze the Adirondack story. Both attempt to give voice to a variety of groups, including state planners, park rangers, well-to-do owners of second homes, middle-class campers, and ordinary Adirondackers. Both explore the conflicts between Adirondackers and the state, though this book covers a later period, from the 1920s through the 1970s. This later timeframe enables us to move past the early conflicts over resource conservation and shed light on the tensions caused by recreational development, which became a state priority after World War I. Jacoby only begins to tap the historical significance of the Adirondacks, for, like so many environmental historians, he turns his attention westward, and moves on to case studies of Yellowstone and the Grand Canyon. Tracing the Adirondack narrative over a longer period provides new insights into changes in recreational trends, environmental politics, and state approaches to conservation.

The conservationist crusade to protect watersheds, control timber extraction, and save wildlife may seem to have little connection to the kind of recreational development examined here.[7] Indeed, the historical literature often pits conservationists against recreation-minded preservationists, as in the Hetch Hetchy controversy of 1913.[8] Nonetheless, during the 1920s, state conservation agencies began to characterize their utilitarian mission as being compatible with recreational development. Building campsites, mountain roads, and other recreational facilities, they believed, promoted healthful exercise, economic growth, *and* rational use of natural resources.

That is not to say, however, that planners' good intentions always led to positive results. People, the built environment, and nature often defied state administrators' attempts to assert control and ensure predictability. It is necessary to emphasize, then, what biologist Daniel Botkin calls "the dynamic rather than the static properties of the Earth and its life-support system" because "life is sustained only by a group of organisms of many species . . . and their environment, making together a network of living and nonliving parts that can maintain the flow of energy and the cycling of chemical elements that, in turn, support life."[9] The energy flows Botkin describes caused campers' waste to show up in drinking water and other undesirable places; and, moreover, the region's topography and weather conditions often frustrated planners' attempts to impose order on the land. The Adirondack region was a space characterized by diversity and change that administrators had to take into account, not a mere subject to be controlled.[10]

Beginning after World War I, administrators, in pursuit of control, applied what David Harvey calls "high modernism" to reshape New York's recreational landscapes. High modernism is the "belief in 'linear progress, absolute truths, and rational planning of ideal social orders' under standardized conditions of knowledge and production. . . . The modernism that resulted was, as a result, 'positivistic, technocratic, and rationalistic' at the same time as it was imposed as the work of an elite avant-garde of planners, artists, architects, critics, and other guardians of high taste."[11] James Scott uses Harvey's notion of high modernism to explain how certain state-directed schemes to improve the human condition have failed. New York's park managers, like Scott's subjects, sought to make the natural and social world more "legible"—that is, simplified so that the state could better monitor and manage it.[12] State planners marshaled the considerable power and resources at their disposal in their effort to make the Adirondack Park comprehensible and comfortable for urban recreation seekers venturing into an unfamiliar environment. Unlike the planners whose failed schemes Scott examines, park administrators achieved their goal—succeeding in opening up the Adirondacks to millions of recreation seekers. Still, forcing legibility on the North Country proved to be a difficult task that required constant negotiation with people and a natural environment that defied simplification. Even success came at a cost.

Park-making from the 1920s through the 1970s stemmed from modernists' and high-modernists' desire to impose order on human and nonhuman nature, and, as such, the Adirondack Park was shaped and reshaped not so much out of an antimodernist impulse as out of an attempt to modernize the landscape. Administrators endeavored to domesticate Adirondack parklands and increasingly over time used science, engineering, and technology to tame the land and its inhabitants, workers, and visitors. They saw little conflict in wielding these means to enable more people to experience nature apart from their urban homes. From the beginning Albany's planners and legislators built the machinery of their centralizing state around its parks. Concurrent with the late-nineteenth- and early-twentieth-century expansion of national bureaucracies, state governments were more frequently and thoroughly intervening in people's everyday lives.[13] In New York and other states, governments began playing a greater role in public health, environmental protection, education, corporate regulation, utility services, and in enforcing vice and decency laws. Managing parks for conservation and recreation was only one of Albany's many new responsibilities. Indeed, recreational development dovetailed with the state's efforts to ensure the public's physical and mental health.

The state's work managing the Adirondack Park was frustrated not only by a complex and unpredictable environment, but also by a growing number of stakeholders invested in the park's future. Among those stakeholders were activists in the environmental movement, which evolved along with the changing natural world. Yet even at the height of the wilderness movement of the 1960s and early 1970s, very few, if any, activists called for a pristine Adirondack Park. To be sure, many environmental activists spoke out against overdevelopment and overuse of Adirondack forests. But quite often they were concerned with wilderness aesthetics, which raised the question: if no one was there to see the wilderness, was it still beautiful? Though Roderick Nash and other scholars have assumed that wilderness connoted the absence of people, even the most dedicated wilderness advocate found a place for human beings in the Adirondack Forest Preserve. Throughout the history of "forever wild," conservation and environmental groups were often at odds with one another and with state land-management agencies over appropriate *uses* of the wilderness. Examining wilderness politics at the state and local levels uncovers the evolving ideas, contradictions, and divisions among nature enthusiasts in the Adirondacks that no doubt had parallels in environmental contests throughout the United States.[14] A variety of social, cultural, educational, and personal factors helped to determine the ways in which people used the land, which in turn shaped their convictions about the natural environment. Mixed land-use patterns in the Adirondacks enabled individuals to appreciate the region's natural features in a variety of ways: as year-round inhabitants, seasonal residents, campers, hikers, hunters, motorists, and skiers. Consequently, recreational development gave people across class lines a stake in the park's future, and their often-competing interests yielded a many-sided, contentious wilderness politics.

Delving into environmental disputes in the Adirondack Park also provides new perspective on local opposition to state regulation and organized environmental advocacy. According to Jacoby, Adirondackers aggressively defended their moral ecology against state conservation policies of the late nineteenth and early twentieth centuries. Carrying the Adirondack story forward reveals that park residents continued to assert their right to use the resources in their communities as they saw fit. We will see, however, that their moral ecology did not remain static. Even though during the middle decades of the twentieth century residents voted against constitutional amendments that enabled construction of large-scale recreational projects in the park, a grudging acceptance of the vital tourism industry emerged. Ultimately, government mismanagement and a worsening regional economy after World War II soured many Adirondackers even more toward the state. By the late 1960s, they had turned to outside developers as their saviors from state supervision and economic malaise. Thus, the home rule and property rights that were so critical to their moral ecology turned out to be flexible principles, and, over time, many locals came to favor interventions by outside capital over those by the state.[15] In this way, residents of New York's North Country would play a leading role in the burgeoning environmental backlash of the 1970s and 1980s.

Clashes over development often resulted from the significant influence that park visitors accumulated over time. The state administrators and local businesses that encouraged tourism made urbanites and suburbanites their chief clientele, and together they forged a strong link between city and park.[16] Just as they did in city parks, urban visitors exerted significant pressure on both the Adirondack environment and park policy, and so planners saw extra-urban parks, which were built to serve the populations of expanding cities, as supplements to urban green spaces. As a result, state management of the Adirondack Park worked hand in hand with city planning, and this collaboration seeped into how conservation officials viewed and administered state parks. Since the state went to great lengths to ensure that masses of urban vacationers felt at home in the Adirondacks, visitors' investment in the park's future gained in strength over time. Even for vacationers who did not own second homes in the Adirondack Park, its landscape came to seem like an extension of home, almost like a huge backyard. The demands of recreation seekers, who sought to protect the amenities found in the modern wilderness playground, clashed with the wishes of developers and many Adirondackers, culminating in intense battles over state land-use policy during the late 1960s and 1970s.

The following panoramic exploration of the Adirondack Park during the twentieth century illuminates the untidy and conflict-ridden history of park-making in the region. During the interwar years, a time

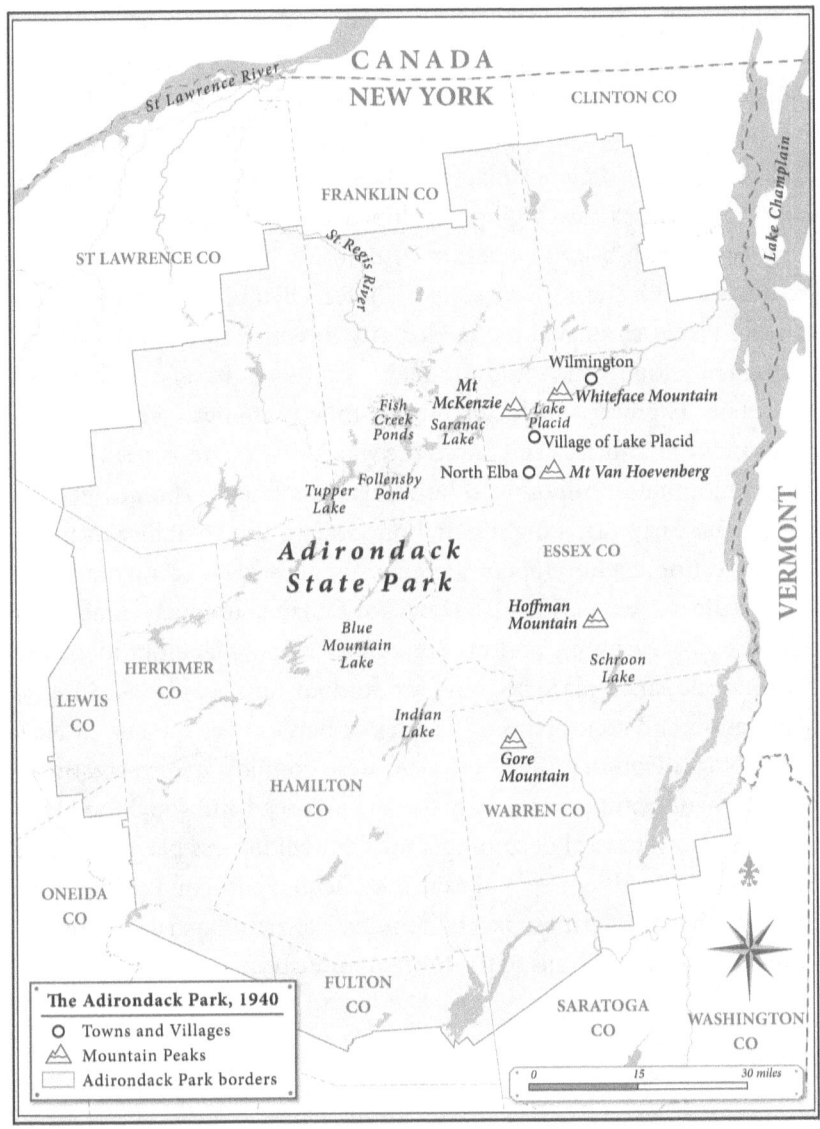

FIGURE 1. The Adirondack Park, 1940. Map by Gerry Krieg of Krieg Mapping.

in which extractive industries in the Adirondacks were on the wane and, simultaneously, urbanization outside New York's North Country yielded a hunger for wilderness among city dwellers, state administrators began offering visitors not only paths to and through the Adirondack Park, but also modern facilities to ensure their healthful play. The Roaring Twenties saw highly developed campsites begin to provide vacationers with a large measure of comfort in the Forest Preserve. As host of both the 1932 and 1980 Winter Olympics, the tiny village of Lake Placid simulated a crowded city in which new infrastructure buttressed indoor and outdoor play. Beginning in 1935, Whiteface Mountain Highway made an almost-mile-high peak accessible to thousands of motorists, and paved the way for two state-run ski centers on the mountain. Proposals to build massive vacation-home enclaves during the early 1970s highlighted once again the flexibility of modern recreation, as developers searched for new ways to turn a profit from Adirondack parklands. New York State's uniquely ambitious role in both developing and managing recreational facilities made the Adirondack Park a place not only for outdoor fun, but also for intense negotiation and contestation—a series of battles over the fate of New York's North Country. The stories of these conflicts over recreational development feature the many players who forged and sought to block the park's path toward becoming a modern wilderness playground, in the process shedding new light on the unequal city-countryside relationship, the long-term impact of deindustrialization on rural communities, and the limitations of the tourism industry.

CHAPTER 1

OLYMPIC TRANSFORMATIONS, PART I

The Re-creation of Recreation and the 1932 Winter Games in Lake Placid

On February 15, 1932, the III Olympic Winter Games in Lake Placid came to end. The village of Lake Placid—located in the town of North Elba, Essex County, the Adirondack Park's High Peaks, and New York State—was now embraced by people from all over the world. During the Winter Games, the village of fewer than three thousand residents had played host to some eighty thousand guests over twelve days, and had witnessed 252 athletes from seventeen countries compete in fourteen events. On that final day, fourteen thousand spectators gathered at Mt. Van Hoevenberg to cheer the bobsledders whizzing by at breakneck speeds.[1]

Several aspects of the Winter Games' concluding bobsled competition were worthy of note. First, the race took place two days *after* the closing ceremonies. The four-man bobsled event was delayed because unseasonably warm weather had prevented safe sledding during the scheduled time. The better-late-than-never bobsled event was just one among many examples of the natural environment shaping the history of play and recreational development in Lake Placid and the Adirondacks. The village's distance from major population centers, scenic beauty, and frigid winters created opportunities for tourism, but at the same time imposed constraints on the local economy. In this instance, winter was not cold enough in a village that was growing increasingly dependent on winter sports for revenue. Despite the best efforts of Olympic organizers to ensure predictability, natural conditions worked independently of human design and planning.

The unpredictability of the weather was not the only challenge Olympic planners faced: they also had to figure out how to accommodate a great many guests. The second striking aspect of the Winter Games' final day was the scale of the bobsled event, which reflected the magnitude of the entire Olympic undertaking. The number of spectators at Mt. Van Hoevenberg was more than four times larger than the village's year-round population. By 1932, Lake Placid was well known in the Northeastern United States as a summer retreat for the well-to-do, who swelled the village's population to 10,000 during the warmer months. Over the three decades preceding the Olympics, the village had also established a reputation as a center for wintertime recreation, thanks largely to the efforts of the Lake Placid Club. In 1905, ten years after the club was founded, the group of affluent winter-sports enthusiasts began hosting a series of ice-skating competitions, skiing meets, hockey games, and other events.[2] The 1932 games represented the escalation of the effort to transform Lake Placid into a winter resort, with the intent to carry on the village's sports tradition well into the future. Entertaining a growing number of recreation seekers required modernizing Lake Placid so that the village could welcome them with comfortable accommodations and state-of-the-art sports facilities.

One of those facilities, the bobsled run, was the third significant feature of the Olympics' final day. Even with a growing winter-sports program, Lake Placid required improvements to its infrastructure in order to play Olympic host. In addition to the bobrun, Lake Placid also gained an outdoor stadium and indoor ice arena. The latter venue was intended to overcome the same weather conditions that would delay the four-man bobsled. Lake Placid also had to accommodate its many guests, which required modernizing housing and transportation in the village. Almost as fast as the speed of the bobsledders racing by was the pace of change occurring in Lake Placid as a result of recreational development.

The final salient note about the bobsled competition was where it *did not* take place. Backed by Albany, the state capital, Olympic organizers had planned to construct the bobrun in the Adirondack Park's state-controlled forestlands. However, wealthy nature enthusiasts, many of whom owned large estates in the Adirondacks, challenged the bobsled proposal. The courts sided with the project's challengers when the justices ruled that building the facility on state lands would violate article VII, section 7, of the state constitution, which prohibited the sale, removal, and destruction of trees in the Forest Preserve. The case challenging the bobrun proposal was the first test of the "forever wild" provision to play out in the courts since its addition to the constitution in 1894, and the justices came down on the side of strict protection. Ultimately, the ruling forced planners to relocate the facility to land that the Lake Placid Club ceded to the state. Nonetheless, the bobsled case signaled the escalation of the protracted struggle over the Adirondack Park's recreational landscape. The bobsled controversy turned out to be one of many environmental and legal contests over the shape of the Adirondack Park during the twentieth century, as an increasingly active state, affluent seasonal residents, and middle-class tourists exercised their growing power to direct land-use policy in the region. The debate over the bobsled run's location revealed the tensions between recreational development and nature protection that would play out during the Adirondack Park's decades-long transformation into a modern wilderness playground.

TOWARD AN OLYMPIC TRANSFORMATION: THE RE-CREATION OF RECREATION IN NEW YORK STATE

The 1932 Lake Placid Olympics took place many decades into a period of accelerating change in New York's recreational landscape. The increasing significance of outdoor play in the Adirondack Park reflected larger social and economic trends during the late nineteenth and early twentieth centuries. First, new transportation routes and technologies made travel both possible and affordable for Americans across class lines. Second, urbanization expanded the pool of tourists eager to leave their home cities and enjoy themselves in the countryside. And third, changes in the workplace gave Americans more time and money to take trips. These three developments posed challenges for governments at every level. A number of politicians and reformers were concerned about what Americans would do with their newfound leisure time and disposable income, and so they sought to provide them with wholesome activities that they believed would benefit individual recreation seekers as well as the common good.

After the Civil War, railroad expansion increased the speed and decreased the cost of travel across the United States, including into the Adirondacks. In 1871, the first rail line into the region stretched from Saratoga Springs to the Warren County hamlet of North Creek. Though the Adirondack Rail Company made the North Country more accessible, traveling from cities continued to be a long ordeal that involved multiple rail lines, ferries, and horse-drawn coaches. To facilitate further travel, rail stops at Ausable Forks and Boonville began operating as additional hubs. Lines soon began crossing the region as well: the Old Forge-Malone line opened in 1891, followed by the Raquette Lake Railway in 1899.[3]

Even with railcars carrying vacationers into the Adirondacks, the state's first priority in protecting the region was not to promote mass recreation, but rather to conserve natural resources. Prior to establishing its management authority over Adirondack forests, the state embraced the concerns of sports hunters and industrialists who feared the exhaustion of the North Country's resources and began a conservation initiative by encouraging large landowners to establish their own private preserves. An 1871 act "for the protection of private parks and grounds, and to encourage the propagation of fish and game," enabled owners to post "no trespassing" signs on their sizable estates and to call for strict punishment of violators: a $25 fine for trespassing and up to thirty days in jail for destruction of property and killing of fish and game. A variety of groups, including the Adirondack League Club of sportsmen and the Lake Placid Club, and affluent individuals, including William West Durant and William G. Rockefeller, established so-called Great Camps in the region (see figure 2).[4]

In 1885, the state expanded its conservation role when the legislature set aside lands in the Catskill and Adirondack regions as the Forest Preserve, largely out of fear that the woodlands' destruction would turn New York into a desolate wasteland. At a time when vanishing wildlife, forestland, farmland, and watersheds revealed the costs of industrialization and urbanization, the Forest, Fish, and Game Commission (1885–1911) and its immediate successor as manager of the Forest Preserve, the Conservation Commission (1911–1927), gave priority to fire control, reforestation, fish and game management, and water-supply monitoring. Progress, according to conservationists, required restraint in order to benefit the public welfare.[5]

With many legislators, industrialists, and nature enthusiasts still unsatisfied with the level of protection given to Adirondack resources,

FIGURE 2. Sagamore Lodge from Sagamore Lake, 1899. Sagamore Lodge was one of the Great Camps that dotted the Adirondacks. It was built during the 1890s by William West Durant, the son of financier and railroad developer Thomas C. Durant. Two years after this picture was taken, William West Durant sold the Sagamore Lodge to Alfred G. Vanderbilt. Courtesy of the Adirondack Experience.

in 1892 the state embarked on a unique experiment in park-making with the creation of the Adirondack Park. The blue line that now set the Adirondack Park apart from the surrounding area on maps encompassed both the state-controlled Forest Preserve and private lands inhabited by about one hundred thousand year-round residents. Two years after the Park's creation, state forestlands received an even stronger safeguard when their protected status was inscribed in article VII, section 7, of the constitution, which stated: "The lands of the State, now owned or hereafter acquired, constituting the Forest Preserve as now fixed by law, shall be forever kept as wild forest lands. They shall not be leased, sold or exchanged, or be taken by any corporation, public or private, nor shall the timber thereon be sold, removed or destroyed."[6] In contrast to places like Central Park, whose creation had required a thorough transformation of the Manhattan landscape and the removal of residents, the Adirondack Park did not initially require much in the way construction, and Adirondackers were allowed to remain in their homes. "Forever wild" ostensibly indicated the absence of extractive industries rather than the presence of recreational facilities, which meant that campers in the Forest Preserve could really rough it if they so chose.

As it turned out, many recreation seekers chose *not* to rough it during their wilderness excursions. Campers in significant numbers had been making their way into the Adirondacks since the 1850s. Many of them were spurred on by Boston minister William H.H. Murray, who earned the nickname "Adirondack" Murray for his popular writings and lectures on the region during the late 1860s and 1870s. Though Murray encouraged campers to play in the rugged outdoors, he recommended camping in "comfortable style," with ample supplies and a guide to lead them through the woods. Tourists who followed Murray's advice to the letter would have spent $125 for travel, board, and "miscellanies" during a month-long camping trip, which cost more than the rates at some fancy hotels. Few families could afford such expenses, but still a great many adventurers left their urban homes for the wild. Stories of unprepared recreation seekers firing their rifles blindly into the woods and engaging in other misadventures made fodder for experienced outdoorsmen and elitists, who dismissed the newcomers as "Murray's Fools."[7]

Rather than run the risk of appearing foolish, affluent and upper-middle-class vacationers had the option of visiting a resort on the Adirondack Park's private lands, where they found many of the conveniences of home. Among the region's most famous hotels was the opulent Prospect House on Blue Mountain Lake, which began welcoming patrons in 1882. As American cities were beginning to electrify, Prospect House won acclaim as the first hotel in the world to provide electric illumination in all of its guest rooms. Leaving one's well-lit room in Prospect House did not necessarily mean a trip into the wild, however, for a bowling alley, shooting gallery, billiard room, and other amenities distinguished the resort as a place for indoor fun.[8] By the early twentieth century, railroads and the hotels and camps they made possible had furthered the transformation of private lands in the Adirondack Park into an amenity-rich playground largely for people of means.

Ironically, urbanites who visited the Adirondacks to escape the stresses of city life were doing so at a time when their standard of living was improving at home. Starting in the 1880s and continuing into the next century, city dwellers began to enjoy the conveniences of modern living: electricity, heat, and indoor plumbing. These amenities improved public health and raised expectations of comfort. Bestowing light on a room with the yank of a cord was easier than lighting a candle; lighting a furnace less labor-intensive than burning wood in a stove; and flushing a toilet much more convenient than using the dreaded, shared outhouse.[9] By the early twentieth century, campers who left their urban homes for Adirondack forestlands were venturing into a more alien setting than the one "Murray's fools" had encountered decades earlier. That is, unless the state invested in infrastructure improvements to provide tourists with a more comfortable recreational environment. As we will see, this would become the mission of state conservationists after World War I.

As tourism became a more popular pastime during the late nineteenth and early twentieth centuries, tensions began to emerge between different groups who staked their claims to the Adirondack wilderness. There were, first, the year-round Adirondack residents, many of whom lived a hardscrabble existence in small towns centered on mining, logging, and tourism. For many Adirondackers, the forest was a site not only for recreation, but also a place for subsistence hunting and gathering to feed their families, and for collecting firewood to heat their homes. Joining them in the woods were the well-to-do nature enthusiasts whose wealth enabled them to reside at Great Camps while cultivating the persona of the sophisticated sportsman. Also venturing into the Adirondacks in greater numbers were middle-class families whose inexperience made them a target of derision for both Park residents and snobbish recreationists.[10] From the perspective of Adirondackers, visitors to the region seemed to be gaining the upper hand in determining land and resource use in the Park.

Indeed, the growing role the state played in protecting the Adirondacks with the creation of the Forest Preserve and Adirondack Park at first had the effect of expanding private ownership and increasing the power of individual landowners in the region. In 1893, sixty private estates encompassed 940,000 acres in the Adirondack Park—much of the land prime hunting and fishing grounds. The Forest Preserve, on the other hand, contained 730,000 acres. With hunting, fishing, and logging curtailed on state lands as well as on private preserves, some locals lashed out. They tore down fences and "no trespassing" signs; took fish, game, and timber from protected forests; and set fires. In one extreme response, an unidentified assailant shot and killed Orrando Dexter, an estate owner in Franklin County.[11]

Also in Franklin County, residents of the town of Brandon challenged William G. Rockefeller's power to restrict the public's use of local resources. Rockefeller had purchased most of the town's land from the owner of a sawmill, and was able to persuade most Brandon families there to sell their properties to him. Not every resident was willing to work with Rockefeller, however. In April 1902, Oliver Lamora ignored the orders of a guard to leave Rockefeller's estate and caught nineteen fish in the St. Regis River. Rockefeller had Lamora prosecuted for trespass. In the case's second appeal, the judge ordered the jury to find Lamora guilty, and he was required to pay the plaintiff's court costs. Victorious, Rockefeller had the Brandon homes torn down to make way for a private wilderness preserve.[12] Adirondackers' uses of the lands and resources where they lived were increasingly restricted, and state policies appeared to favor wealthy vacationers, sportsmen, and second-home owners rather than year-round residents.

In order to promote their own interests, in 1901 a number of seasonal park residents formed the Association for the Protection of the Adirondacks. The group was composed of exceedingly wealthy industrialists and businessmen, including William G. Rockefeller, J. Pierpont Morgan, and Alfred G. Vanderbilt, who lived most of the year in cities outside the Adirondack Park. Two impulses motivated the association's members: a romantic appreciation for natural beauty, coupled with a desire to preserve the isolation, rusticity, and scenery that had inspired them to purchase expansive estates in the Adirondacks.[13] Toward that end, the association urged the state to be vigilant in its conservation of the Forest Preserve and led the fight against attempts to weaken the constitution's "forever wild" provision.

The twentieth-century trend toward mass recreation would test the limits of "forever wild," as changes in the workplace, the proliferation of the automobile, and new government policies encouraged the middle and working classes to play in a variety of green spaces, from urban parks near their homes to national parks. As more white-collar workers enjoyed two-day weekends and paid vacations, and as factory mechanization and increased productivity reduced the number of hours the average blue-collar laborer put in, men and women across the country took advantage of their newfound disposable income and leisure time. Many Americans spent their hard-earned money on automobiles, which conveyed them to the recreational spots of their choice.[14] In 1916, the federal government took two significant steps in promoting travel and outdoor recreation among motorists: the National Park Service Organic Act, which organized the national park system, and the Federal Highway Act, which boosted federal funding for road building.[15] By the beginning of the 1920s more and more Americans were driving their cars to parks to affirm their place in the leisure class.

State governments also did their part to encourage outdoor recreation during the 1910s and 1920s. States as varied as Oregon and Tennessee, Texas and Iowa, committed significant resources to designing and expanding their park systems, and to connecting them with new roads.[16] Since national parks and other open spaces were concentrated in the West, states like New York in the urbanized Northeast had to find creative ways of meeting the demands of an increasingly mobile population. Between 1895 and the early 1920s, the number of state-sanctioned parks in New York grew from a mere six to forty-one. More than a dozen different custodians managed them, including private groups such as the American Scenic and Historic Preservation Society and state bodies such as the Palisades Interstate Park Commission. However, they rarely communicated or coordinated their efforts to address the needs of the state's population as a whole.[17] Consequently, by the 1920s many reformers and politicians began calling for an organized park system that would provide recreation on a massive scale.

In 1921, a new bipartisan reform organization called the New York State Association first proposed the centralization of New York's park system as a means of coordinating recreational planning and development. In a report drafted by secretary Robert Moses, the Association's Committee on the State Park Plan called attention to the need for "a really comprehensive and unified state park plan which will take into consideration the anticipated growth of the state's population and more particularly the growth of cities."[18] The driving force behind the state park movement was reformers' desire to provide New York's urban population with relief from the pressures of work and city living, as well as alternatives to less wholesome leisure-time activities. According to the Second National Conference on State Parks, "The State Park . . . is a growing factor in modern American life, and it is one of the most hopeful, for it is a reaction of the inner instincts of humanity against a wholly new and artificial environment which threatens not only the impairment of its life but the mutilation of its soul." Urban parks were not enough, as Gordon Battle, president of the Parks and Playgrounds Association, explained: "As our growing urban communities become more and more congested the necessity for furnishing fresh air, woodland spaces, and opportunity to enjoy nature becomes more and more urgent and difficult." State parks were supposed to serve as anti-Coney Islands: spaces where healthy outdoor play would reinvigorate visitors' bodies and spirits, and ensure that they remain productive workers upon their return to the city.[19]

Remedying the administrative chaos of New York's collection of parks and satisfying the recreational needs of the growing urban population would require careful planning. In order to meet these challenges, in April 1924 Governor Alfred E. Smith approved the creation of the State Council of Parks, which comprised regional commissions tasked with coordinating park planning and development throughout the state. Parkway construction would be among the council's most significant tasks, as the group endeavored to make state parks accessible not just to the wealthy, but to anyone with a car. Once voters had approved a $15 million bond issue for the improvement of New York's recreational spaces in November 1924, the stage was set for state parks to begin taking on a new character.[20] The State Council of Parks would do much more than open up green spaces, hills, mountains, beaches, and swimming holes to the masses. From Niagara Falls to Jones Beach, from the St. Lawrence River to the Palisades, recreation seekers would find the fruits of the council's labors: modern recreational facilities designed to ensure their healthful and wholesome play in the outdoors.

Although the parks council did not oversee the Adirondack Park, the same developments that encouraged the creation of the council would also have a profound effect on the Adirondacks. In order to adapt to the exigencies of a modernizing nation characterized by growing cities, developing infrastructure, and a mobile population with more leisure time, a new wave of reformers created centralized state apparatuses like the parks council that significantly altered the planning, intended function, and structure of New York's recreational spaces. In their complementary effort to alleviate the pressures of city living, the Conservation Commission and its successor, the Conservation Department (1927–1970), which absorbed the State Council of Parks in 1928, created in the Adirondacks a hybrid environment that melded elaborate recreational infrastructure with rugged nature. The 1932 Olympic games furthered this trend with the addition of new winter-sports facilities on both private and state lands in the Adirondack Park.

THE "MODERN INTERPRETATION" OF "FOREVER WILD": THE THIRD OLYMPIC WINTER GAMES

As recreation was taking on a new character in New York during the late 1920s, Lake Placid officials lobbied for the Olympics in their effort to transform the village into a resort distinguished by modern winter-sports facilities. In 1927, the year before St. Moritz, Switzerland, hosted the second Winter Games, the International Olympic Committee (IOC) inquired with village leaders about whether Lake Placid was up to the challenge of hosting the Olympics. The following year, Godfrey Dewey, a prominent local and founding member of the Lake Placid Club, traveled to several famous winter resorts in Europe, including Chamonix, France, host of the first Winter Olympics, to compare the facilities there to those in Lake Placid. Dewey also attended the St. Moritz Olympics as leader of the U.S. ski team, and studied the town's program and environment. Convinced that Lake Placid was capable of hosting the next Winter Games, he returned home determined to persuade his peers, village leaders, the wider community, and the state government that the winter-sports extravaganza would bring great renown and economic improvement to the Adirondacks.[21]

Dewey stressed the need for Lake Placid to build upon the natural features so instrumental in making the village's reputation as a center for winter sports. In his first speech on the subject, delivered to the Lake Placid Kiwanis Club on March 21, 1928, Dewey touted the village's winter-sports attractions, both natural and human-made. As it had for decades, he promised, a frozen Mirror Lake would serve as the ideal spot for speed-skating competitions. The existing ski jumps at Intervale (like Lake Placid, located within the town of North Elba), along with an expanded system of cross-country trails, could easily accommodate competitive skiers. Though bobsledding and indoor facilities were needed, Dewey was convinced that Lake Placid had the goods: "Lake Placid's quarter century of successful experience in promoting winter sports is its greatest asset in bidding for the Games." The greatest challenge, he conceded, would be housing the thousands of athletes and guests. Two days after his address, the Lake Placid Chamber of Commerce, Kiwanis Club, and Village Board met to plan an official Olympic bid. Attendees formed a temporary committee that included Dewey, the chamber of commerce president, the town of North Elba supervisor, the president of the Bank of Lake Placid, and William Burdet, a local businessman and sportsman.[22]

As the committee's composition indicated, Olympic proponents pursued the games to encourage economic development in Lake Placid and beyond. By the 1920s, logging and iron mining were well into a long period of decline in the Adirondacks, and manufacturing fell on hard times as well. According to the U.S. census, between 1900 and 1920 the number of manufacturing firms in Essex County dropped from 190 to 80.[23] Tourism seemed to be a suitable economic alternative to heavy industry, among few other viable options, in a village nestled in the scenic and rugged High Peaks.

Dewey made explicit his goal of turning Lake Placid into a lucrative winter resort to rival Europe's famous locales. He delivered another noteworthy speech to the chamber of commerce on April 3, 1928, which convinced members to pledge full support for the Olympic bid. Dewey next took his message to local governments and sports clubs in North Elba and neighboring towns. He then went to Albany, where his lobbying paid off when both the assembly and senate passed by unanimous consent a resolution to invite the International Olympic Committee (IOC) to declare Lake Placid host of the 1932 Winter Games. The resolution explained lawmakers' support: "Lake Placid in the Adirondacks offers more complete and adequate facilities and longer and more successful experience in the holding of winter sports than any other community in the United States, and . . . the Olympic winter sports are an inspiration and encouragement to the most wholesome and invigorating type of outdoor winter recreation for the whole people."[24] Consistent with the state's growing role in promoting outdoor play, the legislature's statement emphasized the significance of the Olympics as a step toward providing healthful sport for trained athletes as well as for the masses.

Lawmakers began planning for the construction of Olympic facilities even before the IOC had awarded Lake Placid hosting duties. On January 21, 1929, the Essex County assemblyman Fred L. Porter introduced a bill to authorize construction of a bobsled run on state land. However, the Association for the Protection of the Adirondacks moved to kill the legislation on the grounds that the bobrun would degrade the Forest Preserve and thereby violate the state constitution's "forever wild" amendment. Under pressure from an influential group of New Yorkers, Porter backed down and introduced a new bill that did not specify state lands as the site of construction. On February 21, 1929, Governor Franklin D. Roosevelt signed the bill.[25]

That same month, Lake Placid sent its formal proposal to the IOC for consideration; and while New Yorkers awaited word on the group's decision, the tug-of-war over the bobsled run continued. Porter introduced another bill to approve bobrun construction on state lands, and this time he added a new justification. The winter-sports facility, he asserted, would "induce the people to visit and enjoy the wild forest lands of the state, to stimulate public interest in preserving them for the scenic and recreational purposes for which they were set apart as wild forest lands." In other words, the bobsled run would increase the Forest Preserve's popularity and, in doing so, encourage more people to appreciate state lands as recreational spaces. Mass recreation, he argued, was the key to keeping Adirondack forestlands "forever wild." Time and again throughout the twentieth century, proponents of recreational development justified their projects on the grounds that they would benefit people as well as the natural environment. Porter, Dewey, and their supporters believed that the Forest Preserve needed to be useful, and recreation offered a win-win scenario: the state would prepare the ground for vacationers, who would spend money at their destinations, and in the process learn to cherish nature. Not everyone was convinced, however. The Association for the Protection of the Adirondacks once again protested vigorously, but this time to no avail. On April 9, the legislature passed Porter's bill, and Governor Roosevelt signed it the following week.[26]

Of course, the plan to construct the bobsled run was contingent upon

Lake Placid winning its bid for the Olympics, and on April 11, 1929, the IOC announced that the third Winter Games would take place in the Adirondack village.[27] Dewey successfully made the case for Lake Placid on several fronts: its location within a twelve-hour train ride from New York City; its "unusually dependable climate" that blessed the ground with ample snow and ice; and its decades of experience in conducting a popular winter-sports program. The facilities that made such a program possible, the proposal argued, set Lake Placid apart from its rivals. The natural bounty evident in Lake Placid's iced-over lakes and snow-capped hills had laid the groundwork, and local promoters had prepared the ground for winter sports. Intervale, where the ski jumps were located, "was as perfect as possible for all jumps from 40 meters to 60 meters." Yet it would require some adaptation: "the take-off can be quickly extended about 3 meters which tends to lengthen the average jump about 10 meters." The grandstands at the jump complex seated one thousand, with room to expand seating for an additional two thousand spectators.[28] Whereas Lake Placid's location in the High Peaks had given it the potential to become a winter resort, it was, and would continue to be, the task of determined and organized locals to see that the village fulfilled its promise.

As noted in Lake Placid's proposal, the village would have to undergo significant development to play Olympic host and continue to welcome winter-sports enthusiasts after the games' end. Lake Placid's representatives proposed construction of a new outdoor stadium and indoor ice arena for skating competitions and hockey games. The world's largest bobsled run also was part of the plan, but it was still unclear where it would be built. In addition, Lake Placid's boosters promised the IOC that it had adequate office space, meeting places, and telephone and telegraph facilities for visitors and the press. Housing would be a concern, however. They estimated that the Lake Placid Club, and the hotels, boarding houses, and private homes in the village and surrounding communities could comfortably house up to ten thousand people. Saranac Lake, Lake Placid's neighboring village connected by "excellent railroad and bus facilities," according to the Lake Placid proposal, would bear some of the housing burden.[29]

It was significant that the bidders sold Lake Placid as the ideal spot for the Olympics by touting its existing modern facilities. Advertising itself as a remote wilderness retreat would not have served Lake Placid well in making the case for its suitability as Olympic host. For almost two weeks in February 1932, the Lake Placid area would simulate a small city, the kind that had been popping up and expanding throughout the late nineteenth and early twentieth centuries. As a consequence, the remoteness and wildness that had drawn recreation seekers to state forestlands would not have been effective selling points for Lake Placid. Instead, champions of the Olympics were best served by celebrating the village as a middle ground, with both a climate and natural landscape favorable to winter sports, and human-made communication, transportation, housing, and recreational facilities that improved on nature's gifts. In fact, Lake Placid's boosters promised additional construction that would push the village further in the direction of modern development (see figure 3).

FIGURE 3. Aerial view of Lake Placid, c. 1932. The Conservation Department took this photograph to capture its work in Lake Placid. Though the picture is undated, New York State archivists estimate that it is from about 1932, when the village first hosted the Winter Olympics. Courtesy of the New York State Archives.

Though boosters' words had made a Lake Placid Olympics possible, it would take planning, money, and months of building to see the games come to fruition. On June 4, 1929, Dewey became president of the third Winter Games committee, and Willis Wells, town supervisor of North Elba, became vice president. The promotional work of these local officials seemed to have the intended effect, as excitement for the games spread beyond their ranks. North Elba taxpayers overwhelmingly approved, by a five-to-one margin, a $200,000 bond issue for the Olympic stadium and other expenses.[30] Olympic fever was spreading throughout New York's North Country, but the Association for the Protection of the Adirondacks would put a crimp in organizers' plans when the group challenged the constitutionality of the bobrun site.

Olympic planners had considered a number of sites for the bobsled run. Since Governor Roosevelt had signed two enabling bills, organizers ultimately settled on two different locations. The first was on private land at the base of Mt. Jo, but the Winter Games committee did not consider it an ideal location because it was ten miles away from Lake Placid. Much closer and conveniently situated along a main highway was a spot on the western slope of the Sentinel Range. Unfortunately for the committee, the area was state land, and the Association for the Protection of the Adirondacks stepped in to challenge the legality of building in the Forest Preserve.[31]

The association embraced a strict interpretation of article VII, section 7. Although the "forever wild" provision clearly prohibited logging and other extractive industries in the Forest Preserve, it was unclear whether large-scale recreational projects had a place on state forestlands. Building the bobsled run in the Sentinel Range would require the clearing of some twenty-six hundred trees—did it matter why the trees were cut down, or simply that they were no longer standing? The association argued, "a small encroachment invariably leads to larger encroachments in increasing ratio until the object for which the principle established is lost." Significant development of any kind, the group warned, would lead to further development and eventually render "forever wild" meaningless. For the group's members, an elaborate winter-sports facility had no place in a forest reserved for outdoor exercise and quiet contemplation of one's beautiful surroundings. Besides, bobsledding was an activity enjoyed by the few, they asserted, and a run was not in the public interest.[32] The association's members were loath to admit, however, that one of their goals was to safeguard the Adirondack Park *for* the few—people like them who could afford to reside for a few months every year in a Great Camp. The "forever wild" principle protected public resources, but it also kept the state lands adjoining their estates in a relatively natural state.

Dewey responded with a vigorous defense of his plans. Olympic development, he argued, was consistent with twentieth-century landuse practices and necessary for large-scale recreational uses. He dismissed the association's arguments as "preposterous nonsense," and explained his adherence to what he called "the modern interpretation" of article VII, section 7, "which has meant and will mean so much in developing the public recreational opportunities of the Adirondacks." Dewey argued that Olympic facilities were of a piece with the recreational development that had been occurring throughout New York and other states during the 1920s and early 1930s. He added, "A literal construction of the section [of the constitution] at issue would forbid the cutting of a single tree for whatever purpose and would long since have paralyzed the principal activities which have made the Adirondacks famous as a resort." He called attention to the Conservation Department's construction of fireplaces and other conveniences at state campsites, which seemed to violate the strict interpretation of the constitution.[33] Indeed, state priorities had changed significantly since 1894, when "forever wild" became a governing principle in the Adirondack Forest Preserve. Whereas administrators at first gave priority to resource conservation over recreational development, by 1930 the state had a decade's worth of experience in building modern camping facilities for the masses. According to Dewey, the Olympic effort was consistent with nationwide trends toward mass recreation, and preventing the games from taking place would stifle progress.

Unfortunately for Dewey, in the case *Association for the Protection of the Adirondacks v. Alexander MacDonald, Conservation Commissioner*, the state courts ruled that the measure to permit bobrun construction on state lands was unconstitutional. On January 15, 1930, the state supreme court found in favor of the association, and the attorney general quickly appealed the ruling.[34] The results were the same after the court of appeals heard the case in March. According to the higher court's unanimous decision, "The same plea made for the toboggan slide in winter might be made for the golf course in summer, or for other sports requiring the use or the removal of timber. . . . This plea in behalf of sport is a plea for an open door through which abuses as well as benefits may pass. The Constitution intends to take no more chances with abuses, and, therefore, says the door must be kept shut."[35] New York's highest court agreed with the association's slippery-slope argument and affirmed a strict reading of article VII, section 7. In the process, the justices dealt the Olympic planners, and all developers who may have had their sights set on state forestlands, a major setback. As we will see in the coming chapters, however, the debate over "forever wild" was just beginning.

Nevertheless, the bobsled run would be built, but on private land instead of in the Forest Preserve. The region's status as a state park did not protect all lands embraced by the blue line. Past actions by the state and private landowners had created a park in which certain lands received constitutional protection while others did not. Development thus continued, not necessarily based on the suitability and desirability of the natural environment, but rather based on who owned the land. Disappointed but undeterred, Dewey chose a new bobrun site eight miles from Lake Placid, on land the Lake Placid Club ceded to the state. The Conservation Department began construction of the Mt. Van Hoevenberg Bobsled Run in August 1930. Workers blasted and dug out the tree-lined, rocky surface of the mountain to build the one-and-a-half-mile-long course. To ensure safe use of the run, they installed a gasoline-powered engine that would spray twenty thousand gallons of water through eight thousand feet of underground pipe each day to keep it frozen solid. Laborers also cleared enough parking spaces for two thousand automobiles, and built a clubhouse with a restaurant to make visitors' stay at the run more comfortable (see figure 4).[36] If not for the intervention of the Association for the Protection of the

FIGURE 4. Mt. Van Hoevenberg Bobsled Run finish line, 1932. Onlookers watch as a four-man Olympic bobsled team crosses the finish line at Mt. Van Hoevenberg. Courtesy of the New York State Archives.

Adirondacks and the courts, state workers would have reengineered a mountain in the Forest Preserve for mass recreation.

While many eyes were focused on the conflict over the bobsled run, Olympic development at other locations proceeded apace, even as the effects of the Great Depression afflicted the country. As the bobrun case was making its way through the courts, construction of the Olympic stadium was under way near North Elba High School. Building the stadium required intensive labor and substantial transformation of the landscape. Laborers used gas-driven shovels to excavate almost 152,000 cubic yards of earth. Workers then laid out a quarter-mile track of cinder and gravel, which they proceeded to flood and ice to create a four-hundred-meter speed-skating course. They also cut down 105,000 board feet of Douglas fir to build the grandstand large enough to seat 2,875 spectators. During the games, the stadium would host the opening and closing ceremonies, all speed-skating events, the start and finish of both the eighteen-kilometer cross-country ski race and dog-sled competition, and part of the hockey schedule.[37]

Though the bobsled controversy caused Olympic planners the biggest headaches, they would have to surmount another obstacle to bring the indoor arena to fruition. By December 1930, more than a year into the Depression, the state had spent $500,000 to fund the Olympics, but planners would have to look elsewhere to finance the ice arena after Governor Roosevelt vetoed a new appropriation. The fate of the arena remained uncertain until July 30, 1931, when North Elba voted in favor of a $150,000 bond issue. The arena was completed on January 16, 1932, less than a month before the games opened on February 4. The arena's completion came just in the nick of time, as unseasonably warm and dry weather sent the athletes indoors to practice.[38]

The ice arena allowed, for the first time in the short history of the Winter Olympics, figure skaters, curlers, and hockey players to compete indoors, sheltered from weather that threatened to disrupt the schedule.[39] The second Winter Games in St. Moritz had been suspended for a day and a half as a result of an unexpected thaw, and Godfrey Dewey promoted the indoor arena in order to avoid such a contingency in 1932. As noted in the Olympic committee's official report, "No resort in the world, *outside the larger cities*, can boast of such a building. Never before had any part of a Winter Olympic program been held under a roof." With the arena of brick, steel, and concrete looming over smaller, simpler structures along Lake Placid's Main Street, the resort village now had a sports facility to rival those of large cities. According to Olympic promoters, urban-style development was cause for celebration, for it brought lasting economic benefits and represented progress. They would have liked to see such development take place on state lands as well, but the courts kept major Olympic facilities confined to private lands. The arena, for its part, would serve a useful function during the games. Lake Placid, like St. Moritz four years earlier, experienced an uncharacteristically warm winter, but the arena enabled participants to escape the vicissitudes of nature while playing in a human-made, human-controlled environment. Lake Placid became, thanks to the arena, the only resort in the United States where human-produced ice was available all winter long.[40]

Permanent improvements to Lake Placid's infrastructure were intended to ensure that the Olympics' impact on the region lasted well past February 1932. Consequently, development could not be limited to constructing sports facilities. In anticipation of an unprecedented number of visitors, hotel and cottage owners winterized their summer accommodations. Planners would have to get even more creative, however. In preparation for the games, workers laid five hundred additional feet of tracks at the Lake Placid rail yards to support twenty-five cars that would house five hundred people. To increase accessibility and mobility, the Department of Public Works completed a highway that stretched seven miles from Cascade Road to Mt. Van Hoevenberg. Along this and other area roads, twenty new buses would transport visitors between Olympic venues.[41] For a village with fewer than three thousand permanent residents and a peak population of ten thousand during the summer, accommodating eighty thousand ticketholders, 252 athletes, and a large number of officials and workers proved to be quite a challenge, but one ultimately overcome with considerable planning, effort, and expense. In the process, local and state officials learned what it would take to support a small, temporary city in the heart of the Adirondacks.

The visitors flocking to Lake Placid in unprecedented numbers were drawn by an ambitious publicity campaign befitting an event that took place at a time when modern advertising and communication were ascendant. Department stores displayed signs and distributed fliers. Newsreel crews visited Lake Placid and recorded the progress of Olympic construction for films that went on to be shown before theatrically released motion pictures. Radio listeners, for their part, learned details of the Olympics through a series of programs broadcast by the National Broadcasting Company (NBC) and Columbia Broadcasting System (CBS). Railroads, bus companies, and steamship lines also played a major role in advertising the games; and, along North Country highways, billboards urged motorists to visit the new facilities in Lake Placid.[42] The organizers' publicity committee used every part of the growing consumer culture and modernizing landscape to connect the village to national markets and communication networks.

Of course, the immediate goal was to sell the Olympics, but another priority was to promote Lake Placid so that the village could reap long-term benefits from a carefully planned and expensive undertaking. The snowcapped hills and mountains depicted in the publicity literature would remain at vacationers' disposal every winter. The skiers sliding across the snow-covered ground, the ski jumpers leaping through the air, the skaters gliding across frozen surfaces, and the bobsledders whipping through runs used facilities that would remain long after Olympic guests had left for home. Moreover, the stores that displayed mannequins decked out in sportswear would continue to serve winter-sports enthusiasts for as long as they chose to visit Lake Placid. Local businesses tried to capitalize on the influx of Olympic spectators: for example, Spiegels Service ("America's greatest chemical cleaners") offered $1 dry cleaning through the month of February, and the Whiteface Mountain House served its specialties of buffalo meat and venison ("the usual high class meals") throughout the games.[43] Early signs were encouraging: a drugstore owner in Schroon Lake, located about fifty miles from Lake Placid, reported that on one day in September 1931 "eight or nine" people asked him the best way to reach Lake Placid for the Winter Games.[44] Development for the Olympics, as the games' advocates hoped, would have a lasting impact on Lake Placid's people, reputation, and landscape.

Most significant over the long term were the bobsled run, stadium, and ice arena—permanent improvements to the recreational infrastructure. After the games, the state Conservation Department assumed control over the Mt. Van Hoevenberg bobrun, and the town of North Elba inherited the stadium and arena. The bobsled continued to be a popular attraction after the Olympics, and so the department took steps to improve the facility. In 1938, commercial electric service was installed, and a new public address system enabled announcers to gin up excitement for the races. Almost seventeen thousand visitors took advantage of the run's new features that year. During the early 1940s, the bobrun continued to draw paying visitors to such events as the Governor Lehman Trophy Race and the North American Four-Man. The stadium and arena, for their part, went on to host a variety of activities, including ice-skating competitions, hockey games, and tennis matches.[45]

Lake Placid had to bank on permanent improvements to ensure that the long-term benefits of hosting the Olympics outweighed the short-term costs. As a writer for the *New York Times* commented, "The fact that more than $1,000,000 has been expended in preparation for the Olympics in this mountain village of 3,000 inhabitants seems on the face of it, precarious inflation in these economizing times. But with this investment Lake Placid becomes a world centre of competitive Winter sports." In addition to acquiring prestige and modern sports facilities, however, North Elba also accumulated a $52,468 deficit that would not be paid off until 1973, the year before Lake Placid was awarded its second Winter Games. Nonetheless, the Lake Placid Olympic Committee lobbied informally to bring the Olympics back to the village in 1940, but the games would be canceled as a result of World War II.[46] Despite the costs and complications involved in staging the Winter Games, the allure of the Olympic spotlight and potential economic growth proved too good to pass up in the years and decades to come.

CONCLUSION

The 1932 Olympics acted as a catalyst for recreational development in Lake Placid and its environs. Olympic organizers were pitchmen, selling Lake Placid first to their Adirondack neighbors, then to Albany, and finally to the world. Their most important target, though, was the vacationer looking for a wintertime destination. Recreation seekers were essential to Lake Placid residents' livelihood—and to the entire Adirondack region's economy—as seasonal residents, customers, consumers, spectators, campers, and hunters. The third Winter Games drew visitors with a new stadium, indoor arena, and bobsled run to supplement existing facilities like the Intervale ski jumps. Athletes' bodies and the physical exertion on display at the Games heralded the creation of a more human-centered environment reshaped for mass recreation. Modern facilities and amenities, paid for by local and state taxpayers, were intended to protect people from an unpredictable and sometimes unforgiving natural environment, and to provide comfort and convenience well after the Olympics had come to an end.

Yet the 1932 Winter Games also highlighted the challenge of recreational development in the Adirondacks, especially once resistance from "forever wild" purists surfaced. The Adirondack Park was a peculiar experiment in resource conservation, one that left future generations to grapple with the meaning of a park that contained a patchwork of state-protected forestlands and largely unregulated private lands. At the time of the bobsled controversy, wealthy estate owners had much to lose if the Forest Preserve were opened up to large-scale projects, and so they mobilized to prevent the dilution of "forever wild." However, the recreational development effected by Olympic organizers in Lake Placid and by state conservation agencies at Adirondack campgrounds gave many more people a stake in the Adirondack Park's future. As Dewey pointed out in his rebuttal against the arguments of the Association for the Protection of the Adirondacks, state administrators had been developing Forest Preserve campsites for some time prior to the bobrun controversy. Dewey, it turned out, was not the only champion of the "modern interpretation" of article VII, section 7. In the next chapter, we will see why and how campsite construction proceeded in the penumbra of "forever wild," and the impact state policies had on recreation seekers' camping experiences.

CHAPTER 2

CITIES OF TENTS

Development of Adirondack Campgrounds during the Interwar Years

On its surface, the ruling of the Court of Appeals of New York in *The Association for the Protection of the Adirondacks v. Alexander MacDonald* seemed to destroy any hope of substantial recreational projects on the Adirondack Park's state lands. In its 1930 decision, however, the court also weighed in on the appropriateness of camping facilities in the "forever wild" forest. While shutting the door to bobruns and golf courses, the justices noted, "The Forest Preserve and the Adirondack Park within it are for the reasonable use and benefit of the public. . . . *A very considerable use may be made by campers and others without in any way interfering with this purpose of preserving them as wild forest lands.*"[1] Adirondack woodlands were not to be left untouched and unused. According to the court, particular "reasonable" uses were permissible, and camping, even on a "considerable" scale, was an important form of outdoor recreation that did not endanger state forestlands. The appeals court thus endorsed the state's ongoing effort, begun after World War I, to provide campers with a comfortable environment. State administrators' work toward that end involved building elaborate recreational facilities that would, over time, stretch the meaning of "forever wild."

Despite the constitutional safeguards in place, state-directed development followed the path of mass recreation between the two world wars, as tourism continued the process of supplanting logging and mining as the most important industry in the Adirondacks. Recreation, however, was a fluid and contested concept contingent upon one's assumptions about recreation seekers' needs and wants, as well as one's conception of nature. State planners' notion of recreation as a set of healthful and self-improving activities pursued by urban families in an unthreatening environment profoundly shaped their approach to park-making. Once the features of modern urban living, including electricity and water and sanitation facilities, became more common, the character of the Forest Preserve began to change in ways that mirrored trends outside the park. As such, officials began offering visitors not only paths to and through the Adirondacks, but also modern facilities to ensure their comfort and enjoyment there. The people who effected and were affected by these changes articulated their responses within the context of larger environmental transformations, in particular urbanization. If, as so many people believed, wilderness rested on the opposite end of a spectrum from cities, then new developments on state forestlands pushed the Adirondack Park closer to the city and further away from wilderness. Thanks to the construction of elaborate infrastructure, the park, in both form and function, became closer to home for thousands of new visitors.

The efforts of administrators, workers, and vacationers merged human-built nature with nonhuman nature in a way that, to a significant degree, shielded campers from the hardships of life in the wilderness. The state's high-modernist project entailed making the Adirondack Park legible, comprehensible and controllable for all stakeholders.[2] Through high-modernist planning and recreational development, the wilderness became domesticated: hospitable to great numbers of people who, though they may have been seeking refuge from their urban homes, would be concentrated at Adirondack campgrounds that featured many of the same modern conveniences that made city living more comfortable. Although planners' goal was to enable campers to transcend the inherent ruggedness of nature, unexpected changes in the hybrid environment frequently defied their conceptions and proved that control was elusive. As urban-style problems surfaced at overcrowded campsites, state officials, Adirondackers, and recreation seekers engaged in a protracted struggle over the physical shape of parklands and the place of human beings in nature. Ultimately, many began to question whether building campgrounds was any more compatible with "forever wild" than constructing a bobsled run was.

THE MODERNIZATION OF FOREST PRESERVE CAMPSITES

After the First World War—because of increased road building, the proliferation of the automobile, and American workers' ballooning free time and disposable income—New York State began to assume a more prominent role in developing the Forest Preserve for mass recreation. In taking steps to realize the Adirondack Park's potential as a campground, the Conservation Commission and its successor, the Conservation Department, conceived of their role as promoting the public good on two fronts: first, by offering wholesome play as an alternative to idleness, vice, and commercial amusements; and second, by boosting the North Country's economy. With the formerly dominant logging industry restricted on state forestlands and limited on private lands as a result of the deforestation, severe droughts, and fires that occurred during the early twentieth century, the Adirondack region became increasingly reliant on money from tourists.[3] As we have seen, during the late nineteenth and early twentieth centuries, vacationers learned, sometimes the hard way, that a safe and enjoyable trip into the Adirondacks required more than the clothes on their backs. Ensuring healthful outdoor play by people unaccustomed to life in the wild seemed to require, at least, adequate training and supplies and, at most, elaborate recreational infrastructure. Whereas prior to World War I state planners had defined "forever wild" as the *absence* of extractive industries, during the 1920s and 1930s, the modern interpretation of article VII, section 7, indicated the *presence of elaborate recreational facilities.*

Historian Philip Terrie has compared the Conservation Commission to a chamber of commerce in describing the agency's role as a promoter of tourism. Terrie quotes a passage from the commission's *Ninth Annual Report on the Year 1919* to support his point: "More money is invested in hotel and other properties, more people are employed, more wages are paid, and the annual turnover is greater, [in tourism] than in the entire lumber business, which once figured as the most important activity of the mountains.... As the State-built roads have brought vacationists to the threshold of the Forest Preserve, the commission feels that it is now incumbent upon the State to receive them, make them welcome, and extend the assistance which will make vacations more profitable and enjoyable."[4] Terrie's comparison is apt; indeed, the Conservation Commission frequently communicated and worked with chambers of commerce throughout the Adirondacks. Public campsites were supposed to complement private facilities and promote general economic growth.

More generally during the interwar years, consumer expectations were on the rise, as a variety of institutions providing goods and services promised complete customer satisfaction. In hotels, department stores, movie theaters, restaurants, and even in hospitals, consumers began expecting and demanding the best amenities and the most accommodating personal attention.[5] The burgeoning consumer culture even enabled recreation seekers to purchase a more comfortable camping experience, one that "Adirondack" Murray would have envied. A sign at Moon Hill Camp on Schroon Lake, for instance, promised visitors "every hotel convenience, all athletic and social activities." Even those campers who sought out more rugged terrain made sure to enter the forest well equipped with the latest camping supplies. As reported in the *Chateaugay Record*, "Every day the streets are filled with cars with baskets and tents strapped on the running boards." The *Adirondack Record-Elizabethtown Post* observed, "a party of eight city people, two automobile loads, with skiis [*sic*] ... passed through Elizabethtown on the way to Upper Au Sable Lake to enjoy a camping experience in winter." Among the many items stuffed into these travelers' cars may have been "the greatest portable victrola ever made," or the radios, cots, and chairs that G.S. Franklin & Co. offered consumers for their outdoor excursions.[6]

However, while Terrie contrasts the commission's commercial role with its prescribed role of "protecting a natural resource," in fact, state conservationists did not see promoting local business as being at odds with conserving nature. With the state's help, the seemingly less destructive tourism industry was replacing extractive industries in the Adirondacks. In theory recreational development was one way to protect a natural resource while simultaneously using it.[7] Outdoor recreation, according to administrators, was prudent use of nature, and the government was playing a larger role in promoting such a use. As a consequence, "forever wild" had become compatible with mass recreation even before the appeals court sanctioned campsite construction.

In 1920, as camping was becoming more popular in both the Adirondacks and Catskills, drawing some twenty thousand campers to both regions in the summer months, the state legislature earmarked $2,500 for recreational development of Forest Preserve lands. That year, Conservation Commission workers, who had prior to the allocation of those funds worked solely in fire protection, completed eighteen new camps, built eighty-two fireplaces, and cleared 196 miles of trails in the Adirondacks (as well as two camps, fourteen fireplaces, and sixty-four miles of trails in the Catskills). The commission also began printing "recreation circulars" to encourage safe play in state forests. Commissioners discovered that their modest new developments failed to satisfy demand when new campsites quickly filled up and remained occupied throughout the summer. The Conservation Commission concluded, "The Forest Preserve is the property of the people of the State. It is the right of the Commission which has jurisdiction over the Preserve that it should do everything in its power to make it accessible to the people, and to furnish the necessary facilities for its full and complete enjoyment."[8] Creating a more human-centered environment furnished with modern facilities would become a greater priority in the years ahead.

As Robert Moses and other officials were taking steps to realize a unified state park system during the early to mid-1920s, the Conservation Commission continued to make recreational development a priority in the Adirondacks. In their 1924 annual report, covering the year in which the State Council of Parks was created, the commissioners observed, "It has been difficult to keep pace with the public demand for additional public camp sites."[9] The parks council, for its part, recognized that the sites under its supervision could not meet the overwhelming demand for recreation in New York, and that the Forest Preserve would be an essential space for people interested in outdoor play. Moses, the council's chairman, asserted, "In the further development of the Forest Preserve areas, increased attention should be given to recreation and camping facilities. . . . For that purpose the Forest Preserve must be made more accessible. There must be . . . increased police and ranger supervision within proper limitations."[10]

To meet growing demand, Adirondack Park managers made acquiring suitable land and improving campsites their priorities. For example, state legislators approved two appropriations of $75,000 each in the years 1923 and 1924 to purchase lands in the Lake George area of Warren County. Once tracts had been acquired, the Conservation Commission set out to make them hospitable to a sustained human presence. Accordingly, the commissioners "planned to place first emphasis on providing those camps with proper sanitary facilities and adequate supplies of good drinking water. The study of these problems brought out more forcibly than ever before the necessity of concentrating on large camp sites with several fireplaces instead of many small camp sites containing only one or two fireplaces."[11] Concentration became the rule in the effort to provide a healthful environment for great numbers of campers.

During the early stages of recreational development, however, significant questions regarding the concentration of large crowds of people on undeveloped forestlands began to vex administrators. Planners' main goal was to create a comfortable environment for throngs of visitors who sought a destination wilder than the cities and suburbs where they lived and worked. Twenty million people lived within a little more than a day's journey from the Adirondack Park, and as roads and automobiles proliferated during the 1920s, the Forest Preserve became more accessible.[12] Yet visitors brought with them new questions for administrators: how different should that tourist destination be from vacationers' homes? How wild should the Forest Preserve remain? How much should it be developed? Planners were ambitious in studying and preparing the landscape for visitors. William G. Howard, the assistant superintendent of state forests, promised vacationers predictability: "we can tell you in advance nearly all you want to know except whether the fish will bite and whether you will have fair weather or rain!"[13] Though Howard's statement was an exaggeration, the Conservation Commission did go to great lengths to make the wild nature of the Adirondack Park more manageable for administrators and more domesticated for campers.[14]

Although picnickers, campers, hikers, and hunters intended to stay in the Forest Preserve temporarily, the administrative challenges confronting managers in many ways resembled those faced by city planners of the late nineteenth and early twentieth centuries. Human beings took with them wherever they went, at the very least, hunger, thirst, and the need to release the waste byproducts of the foods and liquids they consumed to satisfy those basic needs. Whenever a large number of people concentrated in a single area, providing clean drinking water and functioning sanitation facilities became the principal public health challenge (see figure 5). Of course, state conservation and health officials needed the help of their clientele. They warned campers to be ever vigilant in maintaining safe, clean, and healthful surroundings. William Howard spread his message through the local newspapers: "1. Assure yourself that your water is pure. . . . 2. Avoid pollution of the water supply. 3. Bury tin cans and any other material that cannot be burned. 4. Keep the camp site neat and clean at all times and be sure to clean up all refuse of dry kindling in the open camp (if there is one) for the next man. 5. Be careful with fire."[15] Individual campers had a part to play in maintaining a salubrious Adirondack environment for themselves and others, but the growing scale of mass recreation required the state to make a substantial investment in ensuring public health.

FIGURE 5. Sacandaga campsite water supply, 1927. This image captures a stone wall and pump house at the state-operated Sacandaga campsite. Beginning in the 1920s, features like this became more common at Forest Preserve campsites. Courtesy of the New York State Archives.

Thus, park planners realized the utility of centralization: assembling as many campsites as possible in a single area, preferably near a major road or highway. In order to render the landscape legible, they applied an urban perspective that emphasized concentration and accessibility. As the Conservation Commission noted, "The concentration of large numbers of campers in limited areas—some of the camp sites will accommodate the population of *a good sized village*—presents problems in sanitary engineering that must be solved satisfactorily for the protection of the public health."[16] Despite the acknowledged difficulties of building functioning sanitation infrastructure for crowds of campers, managers decided to focus on developing large campsites.

One of the most popular sites was located in North Hudson, Essex County, on the Schroon River. Whereas early pioneers would have reached this spot by traveling along the river, 1920s' motorists could take the New York-Montreal highway to the campgrounds. To ease visitors' movement throughout the site, workers cleared brush and built gravel roads. Thirty-one stone fireplaces provided heat for warmth and cooking. Pipes brought water from a nearby spring to faucets distributed throughout the grounds. By the mid-1920s, most state-developed campsites, including the popular Schroon River spot, included at least toilets, tables, benches, and garbage bins for use by campers and picnickers.[17] The growing ease of a wilderness experience that combined the comforts of home and the fun of outdoor play gave new meaning to urbanites' relationship with nature.

The state's effort to domesticate campgrounds enabled campers to focus on enjoying their natural surroundings rather than worry about the ambivalent impact of their interactions with the land. In the process of making recreation seekers feel safe and secure at campgrounds, administrators alienated them from the parts of nature they may have found disagreeable. The water visitors drank came from an invisible source, the waste they flushed disappeared, and the trash they deposited in bins was removed by caretakers. While enjoying the novelty of visiting a park, urban vacationers possessed peace of mind knowing that the state provided them with roads, drinking water, sanitation facilities, and other conveniences to ensure their healthful play in the Forest Preserve. The services that municipal governments provided them at home were now being provided by the state in the spaces where they played.

Maintaining sanitary facilities at campsites, particularly the larger ones, required frequent and careful oversight of the intricate infrastructure. Campsites' water supplies came from a variety of sources. Hearthstone Point Camp, located in the town of Caldwell in Warren County, secured water for almost fifteen hundred weekend visitors from privately owned springs located on an uninhabited watershed. Meanwhile, the five hundred campers who could use Battle Ground Camp on a given day drew their water from the nearby village of Lake George. At Battle Ground, then, the village and campsite were closely linked, the latter relying on the former for sustenance. Wastewater disposal at the camps turned out to be trickier, however, as the growing number of visitors during the 1920s taxed the waste-disposal systems. Caretakers at Hearthstone cleaned the chemical tanks by filling them with one hundred pounds of caustic soda every three to four weeks. The cleaning solvents were dangerous to handle and left behind an offensive ammonia smell that could diminish the appeal of the outdoors. In a sanitation breakdown even more damaging to the assumed healthfulness of nature, at Battle Ground the health department inspector reported the contamination of a spring by seepage from a nearby cesspool and the accumulation of filth and foul odors at six chemical toilets.[18] The sanitation infrastructure at large campsites required frequent attention from state workers left to clean up after hundreds and sometimes thousands of visitors.

Yet even smaller camps posed significant challenges. Eagle Point Camp in Warren County's town of Chester accommodated a maximum of two hundred campers, but securing a safe water supply, even for far fewer visitors than larger sites like Hearthstone Point could host, required substantial engineering. At the southern camp, a motorized centrifugal pump siphoned water from Schroon Lake through a pipe and into a two-thousand-gallon concrete storage tank. Since vacationers used the lake for bathing, boating, and fishing, the water supply was chlorinated so that it remained, in the words of one health inspector, "free from unpurified pollution of animal or human origin." A spring, located at the base of a slope and covered by a protective concrete basin, provided water for the northern camp. At the time of Eagle Point's inspection in July 1929, its two chemical toilets lacked "agitators" to break up solid waste, and trash was accumulating near the southern camp's water storage tank.[19] The smaller camps came under close scrutiny because they posed significant dangers to public health if their infrastructure was not adequately constructed.

Indeed, even though Owens Pond Camp, located near Wilmington in Essex County, had enough space for only fifteen visitors, the facilities there still concerned inspectors. The privy at Owens Pond threatened to pollute a brook located one hundred feet away. The health department inspector advised, "There would seem to be some question as to the desirability of continuing to maintain this camp without developing a more satisfactory water supply and providing better toilet facilities." Instead of building up the campsite, however, the state chose to abandon it.[20] Tiny campsites like the one at Owens Pond did not conform to administrators' standard of legibility. They preferred to concentrate campers in as few sites as possible to make oversight easier, and so they phased out small campsites that harbored public health hazards incommensurate with their size.

The dangers were real. Campers at Connery Pond, near the towns of Wilmington and North Elba, complained of the water's smell and taste there. The state inspector's chemical analysis of the water from the well found a high bacterial count resulting from "a moderate amount of pollution of human or animal intestinal origin." In other words, there were feces in the water, most likely from the septic tank located one hundred and fifty feet away from the well. The inspector deemed the water safe to drink, but recommended that the septic tank be moved at least two hundred feet away from the well.[21] Problems were not always so easily fixed, however. In 1926, inspectors found contaminated water at Crown Point in Essex County, and twelve years later campers once again complained of an unpleasant odor at the campsite. Poorly dug wells were responsible for the earlier pollution while, in the second instance, two privies located on top of tight clay soil near Lake Champlain prevented adequate leaching of waste.[22] Disgusting tastes and odors were nuisances caused by the halting and incomplete domestication process implemented by planners. Consequently, many campers' wilderness experience turned out not to be as safe and enjoyable as state conservationists had promised. As we will see, this disconnect between expectations and lived experience would have a profound effect on wilderness politics in the years and decades to come. The malfunctioning public health infrastructure was more than an administrative problem though. On the ground, caretakers worked diligently to ensure the health of both campers and the natural environment.

Park rangers were run ragged by the overwhelming number of visitors and tasks before them. Their job was to act as mediators between park visitors and their natural surroundings, and in that role they saw first-hand the consequences of mass recreation's imprint on the land. Rangers performed the dirty work of draining, flushing out, and cleaning the often-problematic chemical toilets and urinals. The Conservation Department's *Manual for Caretakers* cautioned that they should keep the caustic soda used in cleaning away from their hands, face, and clothing. The manual also specified that rangers were responsible for registering campers and, if necessary, insisting that parties that stay longer than two weeks move along to a less crowded site. The manual further advised: "caretakers should make a *thorough inspection* of the campsite at least *once each day* to see that" campers maintain their sites properly.[23]

Campsites required day-to-day maintenance: the grounds needed to be cleaned up, fireplaces needed repair, fire notices needed to be posted, and campers' concerns needed to be addressed. As the Conservation Commission pointed out, "it frequently happens that some careless camper goes away from a camp site leaving it in an untidy condition. . . . This means that the cleanup work must be done by the men of our own force in addition to their other work." Rangers were responsible for all of these tasks while making sure to "Take pride in your job and your appearance and in what your State is doing to make out-of-door vacations in the mountains possible for multitudes of people."[24] Whereas prior to the First World War "Murray's fools" and other campers often struggled in their attempts to enjoy a safe and healthful trip into Adirondack forests, recreation seekers now had rangers to look after them. Caretakers served as both law enforcement officers and sanitation workers, and as more campers flocked to the Adirondack Park, their dual role became increasingly taxing.

To improve campground oversight and facilitate caretakers' work, in spring 1925 the Conservation Commission equipped six rangers with motorcycles. Each ranger was stationed at a public campsite in a tent to keep him close to the campers. This set-up "gave him a chance to set an example of a neat camp to other campers, and the fact that he was on the ground enabled him to meet campers as they came to the camp site to assign them suitable locations for pitching their tents and to instruct them as to the rules and regulations for camping and for forest protection."[25] Rangers need not remain fixed to one spot, of course, for they could hop on their bikes and make frequent inspections of surrounding camps and fulfill their duties across a wider range. As they had done in providing roads, drinking water, and sanitation facilities, state planners turned to technology—in this case, motorized transportation—to further the domestication of wild forestlands. A growing and mobile population of recreation seekers required a fast and flexible manager.

These ground-level efforts to improve park administration were soon joined by state-level measures. As part of a larger restructuring of the state bureaucracy, in 1927 Albany reorganized the natural resources management system with the creation of the Conservation Department, and recreation continued to be a top priority for the new agency. The department's Division of Lands and Forests established a new post, Supervisor of Recreational Development, whose job it was to oversee the construction and operation of campsites and other recreational projects in the Forest Preserve. The entire Division of Lands and Forests, its recreational post included, was dedicated to "to developing [the Forest Preserve's] usefulness and availability for the public whose property it is."[26]

Especially noteworthy in this statement were the words *usefulness* and *availability*. The term "forever wild" suggested to some, including the Association for the Protection of the Adirondacks, that the Forest Preserve should remain relatively pristine. But, according to state planners, untouched land had little obvious usefulness for the majority of New Yorkers. By taking steps to realize the recreational potential of the Adirondack Park's forests, mountains, and waterways, administrators believed that they were giving New York taxpayers their money's worth by making these lands useful. Thus, recreation continued to fit neatly into the principles of conservation. Like fire control, reforestation, fish and game management, and water-supply monitoring, recreational development involved rational use of natural resources. The Adirondack Park must be useful, and state conservationists decided that altering the Forest Preserve to a consumable form so that it attracted and accommodated urbanites was the most effective way to do so.

The Conservation Department's efforts bore fruit, as tourists flocked to the Adirondacks' newly improved campgrounds. Campers filled to capacity the twenty large campsites in the Adirondacks and Catskills during the 1927 season. These sites could accommodate a total of fourteen thousand persons at one time, but many visitors had to be turned away. Total registered use of public campsites ballooned from 36,816 in 1927 to 267,886 in 1930, with increases during every intervening year. In 1930, sixteen rangers patrolled twenty-eight large camps covering two hundred acres in both the Adirondacks and Catskills. While the Conservation Department attributed the massive influx of campers to their desire to "get away from the nervous strain of urban existence, and enjoy simple living in the great outdoors," administrators' dedication to improving the Adirondack Park so that it appealed to urban tourists furthered the domestication of the wilderness.[27]

The growing popularity and expansion of Fish Creek Pond campsite reflected in microcosm a significant portion of the Forest Preserve dedicated to mass recreation. In less than a decade after World War I, the site grew from a tiny open camp covering a quarter of an acre to one of the region's most popular attractions, with its one and a half miles of shoreline. Although department planners observed that "far more bona fide campers" who preferred to stay in tents rather than trailers occupied Fish Creek Pond during the 1928 season, even "bona fide campers" enjoyed freshly cleared and graded grounds; roads and a highway bridge; water supply and sanitary facilities; and a dock and two bathhouses. In a telling turn of phrase, the department's annual report described "a city of tents" at the campsite.[28] As a result of state-directed concentration, "forever wild" proved to be a far more flexible principle than the court's 1930 decision on the bobsled run had seemed to indicate.

In another modernizing experiment begun in 1930, the Conservation Department installed electric lights at three popular campsites. In order to enjoy the benefits of electric illumination, one no longer had to stay at a posh hotel like the (now-closed) Prospect House, which had once boasted of being the only resort in the world with electric light in every guest room. At some campsites, light bulbs illuminated each toilet, the area around each pair of toilets, and the interior and exterior of rangers' headquarters. Electric light accentuated existing development and drew campers to the facilities built to ensure their safe outdoor play. Planners asserted that electrification helped them further their dual goal of maintaining the park's wild character and ensuring a safe camping environment: "these improvements proved to be of great benefit to the campers and in no way detracted from the woods atmosphere of the camp sites. It is recommended that similar installations be made wherever possible as rapidly as funds available permit."[29] Such ambitious measures to domesticate state parklands distinguished New York's park development from other regions in the United States. Three years before New York governor Franklin D. Roosevelt entered the White House and five years before his Rural Electrification Administration began bringing electricity to depressed regions in the South—indeed, even before many year-round Adirondack residents had electricity in their homes—electric light brightened caretakers' and campers' paths through the Adirondack Park. Light bulbs' glow helped people find their way through the camps, and also made the land legible to an urban population that came to appreciate the park as a modern wilderness playground.

One camp that featured electric light was Hearthstone Point. The most popular campsite in the Forest Preserve during the 1930 season drew 53,787 campers and picnickers that year. Visitors found there a number of new features in addition to the illuminated toilets and ranger headquarters: a stone wall; a fifty-foot-long pier; additional fireplaces, tables, benches, and chemical latrines; repaired roads; and a wading crib for small children. Taking the Adirondacks' increasingly common designation as a playground literally, the Conservation Department also installed three swings and four seesaws at Hearthstone. The department reported that the playground equipment was popular among small children and recommended that more be placed at campsites and picnic grounds.[30] Development of sites like Hearthstone was altering the character of the Forest Preserve, as campgrounds began featuring modern improvements more commonly found in cities and suburbs, in addition to playground equipment more commonly found in urban parks (see figure 6). Even during the trying times of the Great Depression, recreational development acted as a catalyst for significant changes in the land.

FIGURE 6. Workers laying a culvert at Hearthstone Point Camp, c. 1920. Hearthstone Point Campground would become one of the most popular campsites in the Adirondack Park. Courtesy of the New York State Archives.

Encouraging mass recreation remained a top priority for the state government during the Great Depression. In 1931, while the country was sinking deeper into the economic doldrums, the Conservation Department continued to insist "that in their recreational use to the public lies probably the greatest value of the State lands which constitute the forest preserve." Indeed, with so many out of work recreation became even more important because it helped the jobless and underemployed fill their (often-unwanted) leisure time.[31] Illustrating the effects of hard times on New Yorkers, the department found that the public used parks and campsites in greater numbers but spent less money on concessions. Almost 375,000 campers registered at campsites throughout the Adirondacks and Catskills in 1932. Use was so consistently heavy during the early 1930s that the department enacted a new policy limiting all campers to two-week stays (with chance for renewal) at state-run sites.[32] These numbers indicated administrators' success and also revealed the challenges they faced. The growing human presence in the Forest Preserve showed that their plans were having the intended effect, but at the same time increased pressure threatened campgrounds' natural appeal.

As a consequence, significant human presence began to extend beyond state-constructed campsites. In 1932, forest rangers began issuing permits—a total of 1,464 that summer—to campers in undeveloped locations in the Forest Preserve.[33] The growing number of campers and hikers in the Adirondack Park's wild and hitherto neglected lands revealed a blind spot in planners' conception of recreation. Evidently, developed campsites like Fish Creek Pond and Hearthstone Point had acquainted these wilderness seekers with nature and emboldened them to venture farther into the forest in search of an outdoor experience unsullied by large crowds and urban-style infrastructure. Whether because state-run camps ran out of room or because some campers came to prefer a wilder landscape to play in, more park visitors sought out spaces apart from well-traveled campgrounds. The state's response was to attempt to bring these trailblazers into its purview, and the ranger force was dispatched to police formerly unmonitored trails and lean-tos in the Adirondacks. We are left to wonder, though, how many went uncounted—a significant unanswered question because after World War II wilderness lovers would grow into a far more visible, vocal, and influential constituency dedicated to ensuring that parklands remained as safe, healthful, and attractive as state conservationists promised.

New Yorkers, however, would take a stand en masse to protect the Forest Preserve well before the national environmental movement emerged after World War II. On the ballot in November 1932 was the so-called "recreational amendment," which, if approved, would have modified article VII, section 7, to allow in the Forest Preserve "such recreational facilities as are not inconsistent with its general wild forest character." The provision was the brainchild of state senator Henry E.H. Brereton and assemblyman Fred L. Porter. The latter was the same lawmaker whose attempt to site the Olympic bobsled run on state forestlands had resulted in the court of appeals' affirmation of "forever wild" in *The Association for the Protection of the Adirondacks v. Alexander*

MacDonald. Stung by that defeat, Porter and Brereton promoted their amendment as a compromise designed to ensure legal and reasonable construction of recreational facilities. Porter pointed out the irony that even though the courts had nullified the bobsled legislation, the Conservation Department continued to carry out campsite development "by a succession of illegal acts in flagrant disregard of the fundamental law of the State."[34] The recreational amendment, he argued, would clarify the court's ambiguous endorsement of camping and sanction both campsite improvements and construction of winter-sports facilities.

New York's conservation groups united in opposition to the amendment. The Association for the Protection of the Adirondacks once again stepped up to challenge a reinterpretation of "forever wild." Samuel H. Ordway, the association's president, asserted that the measure would "permit the State, counties and towns to promote any so-called recreational enterprise, such as golf courses, race tracks, aviation fields, [and] rifle ranges." These features made Ordway's list because they would have damaged Adirondack scenery and brought great crowds to state forestlands. Moreover, they did not encourage the kind of genteel and character-building play the blue-blooded members of his group promoted. The state was already providing for New Yorkers' recreational needs, Ordway argued, and loosening the strictures of article VII, section 7, would lead to the degradation of the Forest Preserve. The association and other conservation groups joined with conservation commissioner Henry Morgenthau Jr. and Governor (and presidential candidate) Roosevelt in a campaign against the measure.[35]

The amendment lost in a landslide. The tally in New York City, where voters tended to favor programs to encourage mass recreation, was thirty-five thousand opposed and fewer than ten thousand in favor, while the margin of defeat outside the metropolis was almost eight hundred thousand.[36] The opposition of the Democratic Party and its leader certainly helped to kill the proposal at a time when the worsening Depression swept Republicans out of office. Furthermore, conservation groups' near unanimity on the issue no doubt helped seal the amendment's fate. These facts, however, do not explain why the amendment's defeat at the polls was so resounding. To understand the vote's outcome, we must consider both the state's ongoing recreational-development policies and the nature of the Adirondack landscape in 1932.

At a time when New Yorkers by the hundreds of thousands visited state-run campsites, and unknown thousands ventured into the Forest Preserve for wilderness retreats, people who lived outside the region found what they were looking for in the Adirondack Park. There was enough wild space for nature enthusiasts to find peace and seclusion, and there were plenty of sites for vacationers looking to camp in "comfortable style." Voters' resistance to the Brereton-Porter amendment, therefore, did not represent a backlash against campground development. On the contrary, their vote indicated that the steps state administrators had taken to improve campsite infrastructure were more popular than ever. Changing the constitution to promote recreational development hardly seemed necessary when campgrounds already featured toilets, electric light, playground equipment, and other amenities. By 1932 elaborate recreational infrastructure was widely accepted as being compatible with "forever wild."

After the amendment's slaughter at the polls, campsite improvement in the Forest Preserve continued unabated. During the Depression, recreational development remained important not only because it offered New Yorkers wholesome leisure activities in difficult economic times, but also because it provided jobs. In 1933, the Conservation Department oversaw thirty-two Civilian Conservation Corps (CCC) camps that employed a total of 8,117 men. Eighteen camps were dedicated to forestry work; three to gypsy moth control; three to blister rust control; five to forest-fire control; two to forest improvement and truck-trail construction on State College of Forestry lands; and four to campsite construction and control. Although austerity prevented the state from allocating funds for the development of new campsites, CCC laborers worked to maintain the existing camps. For instance, CCC Camp 9, stationed in Warren County, improved Hearthstone Point Camp, protected Lake George islands against erosion, maintained island campgrounds, and blazed foot trails in the Tongue Mountain Range. By the middle of the decade, sixty-one CCC camps of 190 men each were completing almost all campground maintenance throughout the state. From March 1933 through October 1935, CCC workers completed, among other projects in the Adirondack Forest Preserve, 17 pedestrian bridges, 108 vehicle bridges, 16 latrines and toilets, 5 disposal tanks and cesspools, 40 line feet of sewer lines, and 11 concrete dams.[37] The CCC continued the state's modernizing project begun during the previous decade.

Government-directed improvements were having an appreciable impact on the environment and campers' outdoor experience. Whereas in 1928 the Conservation Department had described a great mass of "bona fide" campers residing in a "city of tents" at Fish Creek Pond, ten years later a writer for *National Geographic* reported that many visitors at the site would not qualify as "bona fide" any longer. He observed, "all manner of roving homes, from sumptuous meal trailers with futuristic furniture and expensive rugs to humble homemade shacks on wheels and travel-stained old tents"; and added, "Some weekends the population of this *city-on-wheels* climbs to around 2,000."[38] Campers' own tools and vehicles indicated their general acceptance of amenity-rich campsites, for the technologies they brought with them reinforced the state project of protecting recreation seekers from the dangers and discomforts inherent in the wilderness experience. A large group concentrated in a city of tents or in a city on wheels was intended to facilitate the state's efforts to maintain a controlled, safe, healthy, and clean environment for vacationers to enjoy. Unfortunately, two months after the *National Geographic* reporter's visit, eleven campers at Fish Creek Pond fell ill.

In August 1938, the worst-case public health scenario came to pass when eleven campers reported symptoms of gastroenteritis. In order to determine the source of contamination, James J. Quinlivan, a doctor from the state health department, visited the camp to speak with the individuals who were suffering from nausea, vomiting, and diarrhea. He learned that cases were confined to six different sites that used the same well pump. Taps providing chlorinated lake water were available, but campers preferred the fresher taste of the well water to the chemically treated lake water. Ironically, visitors' preference for presumably pure natural water over chlorinated water led to them falling ill, thus highlighting the state's role in ensuring public health in an environment many assumed to be inherently salubrious. The water may have been tainted by the many privies located within two hundred feet of the pump, or even by surface pollution. Although inspectors' findings were inconclusive as to the definite cause of the gastroenteritis, Quinlivan learned from Fish Creek Pond's ranger that, in response to frequent complaints, the department had intended to discard the pump that was the likely culprit. Campground infrastructure required constant vigilance to keep it safe, as further demonstrated by another outbreak of gastroenteritis at Fish Creek Pond the following summer.[39] The twin policies of campsite development and concentration seemed to create as many problems as they solved, as campers' attempts to experience domesticated nature were undermined by the urban-style problems that resulted from the breakdown of infrastructure.

In response to evidence and reports of threats to public health, some residents became restless. In 1940, Katherine H. Newbold—a native of Essex, Massachusetts, and a second-home owner with property on Upper Saranac Lake—protested the Conservation Department's plan to develop a campsite at Follensby Pond, located about twenty miles from Fish Creek Pond. She asserted in a letter to the commissioner of public health: "Experience has proved beyond question of doubt that the building of Fish Creek Public Campsite brought untidiness and abuse to the surrounding woods. There is seldom or never to be found any more public grounds that are not littered with rubbish, old food cans, unburied garbage, etc." She went on: "For the sake of public health it seems unwise to increase the proportion of campers in this relatively small area. This is said for the sake of the campers themselves as well as for those who have been in the neighborhood for years. There are thousands of other square miles which are equally as accessible to develop in the Adirondacks. I heartily approve of encouraging this healthy form of life for the public, but not when it will prove a detriment instead of an improvement to health."[40]

From her vantage point as a seasonal Adirondack resident, Newbold witnessed the consequences of the damaging and unhealthful changes taking place near her second home. Concentrating campers at sites like Fish Creek Pond gave the rangers whose job it was to monitor campsites less ground to cover, but at the same time increased the danger of breakdowns in trash collection and sanitation. Much like members of the Association for the Protection of the Adirondacks, Newbold sought to protect her property from the negative effects of mass recreation. She even encouraged the Conservation Department to scatter campsites widely throughout the Forest Preserve, away from her vacation home. But rather than appeal to the abstract principle of "forever wild," she emphasized her lived experience. For second-home owners like Newbold, the Adirondack region's value lay in its fresh air, clean lakes, and scenic beauty—the natural qualities favorable to outdoor recreation. The trash she saw and smelled at campsites was at odds with the ideal outcomes of state plans that promised healthy people playing in a healthful environment. Newbold's letter foreshadowed second-home owners' growing clout and aggressive activism after World War II.

Thanks to government programs and the CCC's efforts, neither the Great Depression nor second-home owners' protests could halt the flood of visitors to the state Forest Preserve; however, uncontrollable natural phenomena could. Months of heavy rains followed a July 1939 to April 1940 drought that, together, curtailed both conservation work and recreation in the Adirondack Park. Approximately 600,000 campers registered in 1940, compared to more than 740,000 the previous year. Recreation in the Forest Preserve took another significant blow the following year when, on April 30, Governor Herbert Lehman followed the Conservation Department's advice and closed all wooded areas throughout the state because of the risk of fire during an exceptionally dry season. The governor lifted the ban on all woodlands outside the Catskills and Adirondacks on May 8, the Catskills the following day, and, finally, the Adirondacks on May 28. While New Yorkers made alternate plans not involving visits to state forests, workers labored to suppress fires that consumed almost 33,000 acres of state-controlled lands. Once the Adirondack Forest Preserve was opened up, registration at public campsites actually increased over the previous year to 640,000, but failed to match the peak attendance of 1939.[41] There were limits to planners' and caretakers' attempts to impose order on an unpredictable environment.

Beginning in 1942, World War II took over for the region's ecological volatility in stemming the tide of tourists into the Adirondacks. The national war effort had a substantial impact on the management and recreational use of the Forest Preserve. The Conservation Department lost personnel when hundreds of employees left to serve in the armed forces. The CCC disbanded and the vital role its workers played in recreational development came to an end. As waging war became the nation's all-consuming mission, the Conservation Department cooperated with the War Production Board to extract resources from the forests and mines in the Adirondacks. Consequently, recorded recreational use dropped significantly. Camping in 1942 sunk to 46 percent of the previous year's total use, and picnicking, bathing, and other activities dropped to a little under 35 percent from the previous year. In addition, trails and lean-tos saw many fewer visitors, and fewer hikers trekked along the woodlands' trails.[42] Gasoline rationing continued to keep visitors away from the Adirondacks in 1943 and until the war's end.[43] Only ten of twenty-nine campsites opened to the public in 1943, and registration dropped to 35,915 campers, a mere 22 percent of the already reduced use of 1942.[44] Although the trying years of World War II prevented many from playing in state forests, the war turned out to be a temporary interruption in the growing popularity of outdoor recreation in the Adirondacks.

CONCLUSION

By the 1920s, when cities more than ever before dominated the landscape, served as homes to more and more Americans, and required more and more resources, the Adirondack region was caught in their gravitational pull. The question for park managers thus became: how could wilder natural regions, whether kept in that condition by insurmountable natural obstacles or by state restrictions on development, keep pace with the modernizing process spearheaded by cities? What would prevent New York's North Country from lagging behind and suffering from a constant economic malaise? One answer was to serve the growing urban population by offering recreational opportunities not available in cities. This would require that rustic spots, to some degree, come to resemble urbanites' homes.

New York State's administrators had neither a static nor a nostalgic conception of the Adirondack Park, as they endeavored to modernize and domesticate public campgrounds for the comfort and convenience of recreation seekers. Constitutional safeguards may have prevented construction of a bobsled run in the Forest Preserve, but they did not halt extensive recreational improvements. The state acted as an agent of development, as did the thousands upon thousands who clamored to the Adirondacks. These visitors made use of the transportation and sanitation networks that made life in the wilderness more convenient, thus taking part in the redefinition of the Adirondack Park into a modern wilderness playground. In the process, countless recreation seekers gained a stake in the future of a park made to look and feel both natural and homey.

However, in the seemingly mundane difficulties of overseeing a state park—overfilled parking lots, campsites covered in litter, malfunctioning toilets—lay profound questions about human beings' relationship to the natural world. If resource management was good, could too much management be bad? In making the Adirondack Park useful, were state planners opening it up to improper uses? Who got to decide what qualified as proper use? Conservationists liked to see themselves as collaborators with nature. Mountains, trees, fields, lakes, and ponds already existed, begging to be used, and state agencies developed the park's landscapes and waterscapes in such a way that thousands of people could enjoy nature's bounty. By making the Adirondacks legible to both park managers and urban recreation seekers, planners believed they were improving the land. Nature was not always a willing collaborator, however. On a daily basis, visitors' waste followed natural energy flows through the park's ecosystems and turned up in unwanted places like springs, wells, and lakes. Less frequently but no less destructively, fire and drought hindered forest management and outdoor play. And so, in an effort to make the Adirondack Park comfortable for urban visitors, planners looked for new ways to circumvent and transcend the often intractable and unpredictable natural setting where people went to play.

While the state went to great lengths to create a legible environment, however, recreation seekers and residents found ways to elude and protest state oversight. Inspired by their taste of domesticated nature, thousands went on to blaze their own paths into the wilderness, a practice that the state at first ignored, and then began to observe in its peripheral vision. As the interwar years were coming to a close, some second-home owners spoke out against development that they believed threatened the peacefulness and healthfulness of their properties. Just as year-round residents were willing to defend the park they called home, so were visitors prepared to protect the recreational spaces they came to cherish—and their voices would grow louder in the years after World War II.

Before we get to the postwar years, however, we turn in the next chapter to controversial developments that took place in the Forest Preserve outside public campgrounds. Although voters overwhelmingly defeated the recreational amendment at the polls in November 1932, New Yorkers were not opposed to all modifications to the "forever wild" provision. Five years earlier, a majority of voters had approved a constitutional amendment that empowered the state to build a road up Whiteface Mountain; and fourteen years later, they voted in favor of a state-run ski center. Recreation was an elastic concept whose flexibility could be seen reflected in the ever-changing and increasingly developed landscape, as well as in the shifting terms of the debate over "forever wild."

CHAPTER 3

A MOUNTAIN TO CLIMB

The Transformation of Whiteface Mountain
and the Future of the Adirondacks,
1925–1945

Beginning in July 1935, drivers who made a left turn at the town of Wilmington's only stop sign would see a sign pointing the way to the eight-mile-long Whiteface Mountain Memorial Highway (see figure 7). Continuing along the road through the Essex County town, they eventually came across a tollhouse and visitor's center. Here began the trip up Whiteface, New York's fifth tallest peak. At twenty-seven hundred feet up the mountain, the first impressive view at Union Falls Overlook revealed to the north Taylor Pond, extensive woodlands, and mountains. At the road's 6.7-mile mark, the village of Lake Placid and the High Peaks were unveiled through motorists' windows. After traveling another mile, they arrived at the end of the road, where they parked their cars in a paved lot. Beginning in summer 1938, motorists had the option of entering Whiteface Castle, a granite structure furnished with a gift shop, cafe, and bathrooms. From there, they had the choice of climbing a twenty-six-story staircase or taking an elevator to the summit. Once at the top, 4,827 feet above sea level, automobile- and elevator-aided climbers enjoyed, from

FIGURE 7. Whiteface Memorial Highway, c. 1940. The eight-mile-long highway snakes its way up the rugged peak. Courtesy of the Adirondack Experience.

the comfort of a shelter house, a panoramic view of the Adirondack Park's natural wonders. The view was motorists' reward for navigating hairpin turns along the twisting, icy, windswept road.[1] In addition to taking advantage of an expensive and controversial modern feature that made the region's stunning scenery viewable, these travelers were taking part in the re-creation of Whiteface into a mountainscape for mass recreation. Whiteface Mountain Memorial Highway would be more than a road up a mountainside; it would be a path to a different future in the Adirondacks.

The hybrid Whiteface mountainscape, where elaborate human-built structures facilitated visitors' engagement with nonhuman nature, was a manifestation of state planners' flexible conception of modern recreation. Recreational development, as we have seen, involved the building of a stadium, arena, and bobsled run in Lake Placid for the Olympics; and roads, fireplaces, picnic areas, electric lights, and sanitation infrastructure at state campsites. Yet recreation within a modernizing Adirondack Park was a concept fluid enough also to allow for the construction of Whiteface Memorial Highway—an ambitious project that paved the way for the elevator, castle, shelter house, and, after World War II, two ski centers. With these facilities the state augmented its investment in the North Country's tourist economy, thereby enabling thousands to enjoy a relatively inaccessible mountainscape while maintaining its close alliance with the region's businesspeople.

The highway did not go uncontested, however. From 1925 to 1927, proponents shepherded the constitutional amendment to authorize the project through two sessions of the state legislature and a statewide vote, and they met resistance from nature enthusiasts at every step. Preservationists and seasonal Adirondack residents in the urban upper class sought to prevent the dilution of "forever wild" and save nature from modern development. The majority of Adirondack voters also rejected Whiteface Memorial Highway, but likely for different reasons than their more affluent counterparts. For year-round Adirondackers, the mountain road was yet another encroachment by the state that would bring even more vacationers to their home turf. Much to opponents' chagrin, however, the project would go ahead and force a new reckoning of "forever wild."

The divide between proponents and opponents of development on Whiteface was not an impermeable wall, however. Self-described conservationists could be either foes of large-scale construction projects or agents of recreational development, depending upon their conception of appropriate land-use practices and their perception of the environment. Positions shifted in unexpected ways as both Whiteface's physical terrain and the meanings various groups invested in the peak changed. Once the mountain and its environs underwent development and they shifted away from wilderness closer to a domesticated environment, even some "forever wild" purists began to accept the mountain's transformation and advocate for additional recreational projects.

CONSTRUCTING WHITEFACE HIGHWAY

In 1925, the New York State legislature adopted an act proposing an amendment to article VII, section 7, of the state constitution to permit construction of a highway through the Forest Preserve, from Wilmington to the top of Whiteface Mountain. The proposed highway was to serve two purposes: as a memorial to World War I veterans and as a scenic road intended to draw tourists and thereby boost the local economy. The act's success in 1925 was just the first step, as the legislature would have to approve the measure again in the next session, which lawmakers did in 1927. Finally, New Yorkers would have to vote on amendment 7. With the November 1927 vote approaching, both supporters and opponents of the highway responded to the proposal.

The road's most vocal advocates were New Yorkers who championed modern recreation's potential to boost the Adirondack region's economy. The chambers of commerce of Saranac Lake, Lake Placid, and Wilmington endorsed the proposal and formed a committee to work with other chambers of commerce and the American Legion in championing the mountain road. These groups had a powerful ally in governor Al Smith, who also favored the project.[2] Once again, the state and commercial groups worked together to promote tourism in the Adirondacks. The Adirondack Civic League joined them in their efforts, calling the highway "an act of patriotism as well as progress."[3] Progress meant, as it so often did during the interwar years, recreational development for the enjoyment of motorists and the economic benefit of Adirondackers. Whiteface Memorial Highway would not be a thoroughfare connecting towns or cities, nor would it facilitate the shipment of goods. Rather, the highway, like the Blue Ridge Parkway and other scenic roads that were built during the Great Depression, served a purely recreational purpose.[4] They provided temporary construction jobs and, over the long term, new options for recreation seekers with more time on their hands. The Whiteface road would be both the traffic artery and the scenic destination—one of visitors' many destinations, development advocates hoped, among local hotels, shops, and restaurants.

Although Whiteface Memorial Highway was, of course, a matter of great concern to locals who presumably stood to gain either politically or financially, supporters from outside the region expressed strong opinions on a matter affecting a prominent feature of the Adirondack State Park. By this time, thirty-five years after the creation of the park and almost a decade into the state's campsite-development program, New Yorkers across the state were deeply invested in the Adirondacks. Jerome V. Jerome, chairman of the Long Island Chamber of Commerce Conservation Committee, urged New Yorkers to vote yes on amendment 7. "As a conservationist I commend it emphatically as a splendid forward step," he wrote. "Whiteface will convert millions into nature lovers who heretofore have been indifferent, or even hostile, to conservation that makes for the economic betterment of our commonwealth and promotes the happiness and well being of all our citizens." The rhetoric of democratization, opening Whiteface to the middle and working classes, was common in highway advocates' statements, just as it was in the arguments for recreational improvements in general. The *Lake Placid News* endorsed the highway in similar terms: "the Whiteface road will simply make the enjoyment of one of the greatest parts of that great heritage possible for thousands instead of for the few hundreds who are now able to get to the top of the rock-scarred peak."[5]

Developing Whiteface, making it more accessible, would allow motorists of any social class to enjoy the Adirondacks and potentially inspire a new nature appreciation among more people.

As Jerome's remarks suggest, however, battle lines over the highway were not clearly drawn. Many conservationists saw value in the road. Like Jerome, they believed in the potential of the Adirondacks' scenery as viewed from Whiteface's summit to create more nature enthusiasts. Indeed, the Izaak Walton League and the New York Conservation Association both asserted that building the road would allow for appropriate use of the mountain and appreciation of its surrounding scenery. The editors of the *Lake Placid News* summed up this rationale with two points in particular: "Of the 44 peaks in the Adirondack Park over 4,000 feet in height Whiteface Mountain is the most frequently climbed." Whiteface was already a popular site experiencing frequent use, and so, the argument went, it would be wise to make that use easier, more efficient. Thus, constructing the highway, like building up public campgrounds, fit snugly into the state's utilitarian conservation mission. The editors added, "Whiteface Mountain is the only high peak of the Adirondacks naturally adapted to the construction of an automobile road."[6] This statement rested on the assumption that a road *should* be built up a mountain, and so the logical question that followed was *well, which mountain?* From this perspective, New York State's geological history had created a mountain suitable for the construction of a road that could bring motorists near the top. Without the presence of those automobile-borne tourists, highway advocates argued, the mountainscape could not reach its promise as a revenue-generating recreational site.

By contrast, the road's detractors rejected the notion that a highway should be built on a mountain, arguing instead that climbing should be a physically demanding task. A *New York Times* editorial expressed the concern that driving up a mountain required no appreciation for the hardship of strenuous outdoor exercise. The editors pointed out that there already existed a trail up the peak. The footpath was only for those in "first rate condition," they wrote. "If he is fresh from the sluggish life of the city he had better save some of his energy for the last mile. . . .

But the effort will be more than rewarded by the magnificent view from the top." The *Times* editorial page drew a sharp distinction between the rugged life in nature and the soft life in the city. Constructing Whiteface Memorial Highway, the editors asserted, would be committing the sin of bringing an urban feature that catered to weak city dwellers into the wilderness. Ironically, these residents of the New York metropolitan area derided city living, in the process impugning the strength and masculinity of their fellow urbanites who could not endure a mountain climb. The editors went on: "Motor roads . . . enable us to get into the wilds with ease and dispatch. But we don't want too much ease, and perhaps the purposes of recreation could be best promoted with a little less dispatch." Relying on the centuries-old idea that recreation should be self-improving and physically invigorating, the authors argued that the view from atop Whiteface must be a reward for those willing to exert themselves climbing the peak.[7] Whiteface Memorial Highway would be a road too far.

The *Times* was joined in opposition by preservationists who feared that the road and the automobiles and motorists it brought would mar the beautiful mountainscape. Maxwell Steinhardt, a New York City lawyer active in conservation issues, wrote, "To build a motor road up the side of the mountain would unquestionably destroy much of that beauty; and with the motor road would come the inevitable ugly small hotels at the peak and the uglier poster advertisements along the route up."[8] Steinhardt was another urbanite who wanted to maintain a clear separation between city and wilderness—he lived and worked in the former, played in the latter. His words echoed the concerns of the Association for the Protection of the Adirondacks, whose members feared that the Whiteface project would violate the original intent of article VII, section 7, and establish a precedent for further development and commercialization.[9] A road was more than a road, for not only would it bring a physical transformation of the mountainscape, which was significant enough, but also a redefinition of people's perceptions of the mountain. If the amendment passed, the presence of large numbers of people who did not resemble the blue-blooded members of the association would become an accepted feature of the reconfigured mountain.

Modern facilities that encouraged mass recreation, highway opponents argued, were incompatible with a mountain wilderness.

After much debate and despite strident resistance to the highway, amendment 7 passed in November 1927. Sixty-four percent of voters approved the measure by a margin of 1,082,864 to 602,395. Willis Wells, town supervisor of North Elba, chairman of the Whiteface Memorial Association, and future vice-president of the III Winter Games committee, credited the work of 88 organizations and 123 New York State newspapers in ensuring the amendment's success. What he did not mention, however, was that New York City voted overwhelmingly in favor of the amendment, while those living in the Adirondacks opposed it by more than twenty-three thousand votes.[10]

Although advocates promoted modern improvements as a boon to the North Country's economy, such development appeared to favor vacationers more than Adirondackers. As seen with the aggressive defense of their moral ecology against early state conservation policies, Adirondackers were fiercely protective of their homes. Beyond local governments and chambers of commerce there existed a palpable resentment toward projects that would give the state and outsiders another foothold in the region. As more and more visitors came to enjoy the Adirondack Park as a recreational space, however, the relatively small number of Adirondackers lost the numbers game. The *New York Times*, Maxwell Steinhardt, and the Association for the Protection of the Adirondacks did not represent a majority of New York City voters who, unlike highway opponents, valued the Adirondack Park as a space for mass recreation. Ironically, Frederick Stuart Greene, the state superintendent of public works, who designed the highway, had once counted himself among the road's detractors. He had given voice to Adirondackers' skepticism when he wrote the state legislature to deride the proposal's feasibility and utility: "to ask the taxpayers of this State to spend $1,853,000 for a luxury which will benefit so few for such a short time each year . . . is . . . the most extravagant and useless proposition for which your honorable bodies have been asked to make an appropriation."[11] Greene, as we will see, was not the only one whose perspective on Whiteface and other development projects would evolve over time.

After voters adopted the amendment, the legislature included in the 1928 budget a $15,000 appropriation for surveying potential routes for the road up Whiteface, and on April 9, 1928, lawmakers passed the Whiteface Act authorizing construction. The act also established the temporary three-member Whiteface Mountain Highway Commission, which was in charge of establishing tolls, and appropriated $150,000 to begin construction. Work cutting trees to clear the road's path began on Christmas Day, 1931. Workers proceeded to fell thousands of trees, and the Conservation Department kept the spruce for use in campsite construction.[12] The development of Whiteface facilitated the ongoing construction of state-run campsites.

Removing trees was only the beginning of Whiteface Mountain's transformation. Constructing the eight-mile-long, twenty-foot-wide highway, along with retaining walls, shoulders, and parking spaces, required backbreaking work in harsh conditions. A reporter for the *Lake Placid News* described the construction scene on the mountain: "The grinding of the shovels, the clatter of falling rock go on constantly, punctuated by occasional blasts of dynamite, except when clouds blanket the mountain, slowing and sometimes halting work." Another observer detailed workers' efforts as they neared Whiteface's summit: "powder gangs [were] blasting out the solid rock ahead of them. Three air compressors feeding nine drills and two others furnishing power to five drills each, are being used in an effort to keep pace with the trucks which carry the rock and dirt away." Laborers blasted through solid granite and laid down the macadam road. Their deepest excavations reached thirty feet into the mountain and yielded the raw materials used to construct the retaining walls. In addition, they built a tollbooth at the highway's beginning and another at its end, and the necessary electrical and telephone lines. A project that required intensive labor and substantial engineering on a frosty, windswept mountain predictably faced delays. Harsh winter weather prevented the road from opening as planned in September 1934.[13] As they so often do, attempts to transcend the ruggedness of the natural environment hit a snag.

Finally, on July 20, 1935, Whiteface Memorial Highway opened to much fanfare in the Adirondacks. Greene took the opportunity of the road's opening to bash opponents of the project he designed (and had once opposed): "We have been subjected from the beginning to the most ignorant, stupid, vicious and even, I am sorry to say, dishonest opposition. . . . The propaganda began, put out chiefly by professional conservationists[:] . . . those who love to enjoy nature from a swivel chair."[14] Greene's characterization of highway critics was easy to make in light of their privileged social status and apparent elitism. Greene turned on its head their argument that the road catered to soft people who could not handle a mountain climb. Conservationists, as Greene saw them, sat indoors all day and fought to keep the natural environment so pristine that it would not benefit the majority of recreation seekers. The debate over Whiteface turned into a contest of masculinity wherein both sides insisted that their vision for the mountain, and the natural environment in general, best promoted healthy manliness.

Following Greene's vituperation, benign history and nostalgia ushered in a new phase in the development of Whiteface Mountain. The first vehicle to make the climb up the highway was a seventy-five-year-old stagecoach driven by a descendant of the first man to build a home in the nearby village of Lake Placid. In a nod to the road's commercial potential, the coach carried a trunk that had once contained the belongings of the first paying visitor to the resort village. That stagecoach was soon followed by thousands of horseless carriages, as more than one thousand motorists drove up the mountain the weekend of the road's opening. Within a month, more than 17,000 persons in 5,647 automobiles had made the trip. During that same time, only one stalwart adventurer used his legs to climb up, and one man rode a horse.[15] Within a year, more than 61,000 sightseers had wheeled up Whiteface. Evidently, as the *New York Times* put it, "Legs . . . are somewhat less fashionable today as popular means of locomotion."[16]

As the mountainscape changed, so did visitors' recreational experience on Whiteface. The *Plattsburgh Daily Press*, which supported the highway, used a telling turn of phrase to capture the incongruity of a road twisting up a mountainside: "Tomorrow old Whiteface is likely to wake up in the morning thinking that during the night she had been dropped down to Broadway and Forty-second street, for the building of the highway up the side of the mountain thousands who would never have thought of climbing up to the summit have become mountain conscious and will be most happy to ride up at their ease."[17] The city had come to the wilderness and, in doing so, made the wilderness accessible to the city. Whereas prior to the road's completion a nature lover could reach the mountain's summit only with a rigorous climb, by the mid-1930s climbers could enjoy a relatively leisurely, scenic journey up Whiteface by automobile. They could, thanks to the state's new dedication to recreational development and hundreds of workers' labor, transcend the inherent ruggedness that had made the mountainscape inaccessible to them in the past.

Building a road up Whiteface and increasing its accessibility, as both the proponents and opponents of the highway expected, had an impact on the mountainscape beyond the road's path. Anticipating the growth of Whiteface's popularity, in summer 1932 the Conservation Department opened a new campsite alongside the road. The campground covered fifteen acres and offered a magnificent view of the Au Sable Valley. The site comfortably accommodated fifteen hundred campers who could take advantage of the stone fireplaces, electric lights, and water faucets supplied by the state. Since the highway was intended to facilitate visitors' engagement with nature, building a modern campsite like the ones at Fish Creek Pond and Hearthstone Point furthered that goal. But work did not stop there. In 1934, a Civilian Conservation Corps camp opened at Whiteface. To accommodate the mountain's new visitors, CCC workers constructed, between May 1934 and April 1935, thirty-two miles of ski trails and bridle paths and seventy miles of foot trails.[18] Though Whiteface may not literally have come to resemble a busy New York City thoroughfare, the highway did open the door to additional improvements that made a mountain visit possible and comfortable for people from any part of the state.

Some of the proposed improvements were particularly ambitious. In May 1935, Governor Herbert Lehman vetoed a bill that would have issued $1 million in bonds to build a new trail from the end of the highway to a tunnel leading to the center of Whiteface. The tunnel, dug into the mountain itself, would lead to an elevator that would carry visitors to the summit, from where they could view Adirondack scenery from a new shelter house. Greene took up the cause for these facilities and asserted that walking along the windy, three-hundred-foot-long gravel path from the road's end to the mountain's summit was too dangerous.[19] Foot trails had once been the only paths up the mountain, and yet by the mid-1930s, after campsites and the highway had made possible a more comfortable outdoor experience, some believed that walking the last three hundred feet up an almost-mile-high peak was too perilous.

Opponents of the scenic highway who worried that its construction would set a precedent for further development of Forest Preserve lands saw their worst fears realized. Raymond H. Torrey, the secretary of the Association for the Protection of the Adirondacks, wrote the *New York Times* on behalf of his organization: "the shelter and elevator shaft will conventionalize the summit in a manner that will destroy its former majestic wilderness character . . . It will make . . . the ascent of Whiteface by automobile and elevator about as much of a trip as a subway ride in New York City."[20] For Torrey and other likeminded preservationists, the Adirondack Park and New York City rested on opposite ends of a spectrum, and introducing features from the latter to the former irrevocably tarnished the park's character. Unlike the *Plattsburgh Daily Press*, which had compared Whiteface to New York City to celebrate the highway's democratizing effect, Torrey's urban analogy contrasted the humdrum ordinariness of city life with the unique sublimity of Whiteface. A beautiful natural feature, according to Torrey, was supposed to be remote, open to the few, presumably to people like the well-heeled association members. Only then could it remain in a wild state. Unfortunately for opponents of the elevator and shelter house, the state legislature and governor eventually approved funds for these projects.[21] By summer 1938, visitors to Whiteface could enjoy a meal at Whiteface Castle, then take the elevator to the mountaintop, and relax at the shelter house before driving back down the road (see figure 8). Whiteface Mountain began to look a bit more like Broadway and Forty-Second Street, but instead of watching a theatrical performance, visitors took in a show put on by the Adirondacks' spectacular scenery.

In essence, Whiteface Memorial Highway's proponents celebrated

FIGURE 8. Whiteface Memorial Highway and summit, c. 1940. The end of the highway at the mountain's summit. The Whiteface Castle and shelter house are visible. Courtesy of the Adirondack Experience.

the effect that the mountainscape's natural features would have on visitors and locals, while the road's opponents lamented the impact visitors would have on the mountain. A writer for the *New York Times* summed up the highway's significance: "Time was when [the Adirondack region's] remote, quiet valleys with their rushing brooks and placid, forest-encircled lakes, the blazed trails to the high peaks, were the spiritual possession of the few—and those few chiefly the well-to-do. More and more rapidly they are becoming the recreation grounds of the people as a whole."[22] Thanks to the transformation of public campgrounds and Whiteface Mountain, the Adirondack Park had been reconfigured as a site that welcomed not only patrons of expensive hotels and wealthy second-home owners, but also the middle and working classes. Recreation seekers enjoyed not only greater access, but also greater comfort on public lands. Reducing the hardships of the wilderness experience became the mission of state agencies, and the outcome was a much-altered park that would see further domestication in the decades ahead. "Forever wild" proved to be a flexible concept when weighed against the growing significance of modern recreational development. The determination of development proponents to transform the Adirondacks in the interest of mass recreation would eventually furnish the people flocking to Whiteface with another method of descending the mountain: on skis.

TO SKI OR NOT TO SKI: TOWARD A "SKIER'S PARADISE" ON WHITEFACE MOUNTAIN

In February 1935, five months before the opening of Whiteface Memorial Highway, conservation commissioner Lithgow Osborne wrote a piece in the *New York Times* in which he marveled at the rising popularity of skiing: "the growth of skiing in the Northeastern section of the United States has been little short of amazing. The old idea that skiing was only for the youthful and the extremely active person has gone by the board. . . . Various means, ranging from automobile buses to heavy ropes arranged endless chain fashion, have been inaugurated to transport skiers to the tops of runs."[23] Improvements in the technology of transportation and sporting equipment created new opportunities for more people to take part in a recreational activity that combined the thrill of sliding downhill and the comforts of a heated ski lodge.

Just as Whiteface Memorial Highway enticed thousands of motorists into becoming mountain climbers, new ski centers throughout the Adirondacks drew countless enthusiasts who let gravity pull them down snow-covered slopes after human-built devices had carried them up. The mountain climb used to be the hard part, but as rope tows, T-bar lifts, chairlifts, and gondolas did the work formerly done by legs, racing downhill became a more enjoyable and repeatable activity. However, recreationists' growing enthusiasm for skiing posed a new challenge to state administrators like Osborne: how could they cater to a growing segment of the public whose sport of choice required substantial infrastructure improvements on "forever wild" lands? Planners saw at Whiteface an opportunity to exploit further the mountain's malleability as established by construction of the highway, and to build on existing infrastructure with a modern ski center.

By the time Whiteface Memorial Highway opened in summer 1935, downhillers had a growing set of options on privately owned and locally controlled lands in the Adirondack Park. Beginning in winter 1934, so-called snow trains, complete with coaches, sleeper cars, and cars dedicated to selling and renting out equipment, carried hundreds of skiing enthusiasts to Warren County's village of North Creek, where the slopes of Gore Mountain beckoned. Tupper Lake's Sugar Loaf, Speculator's Oak Mountain, and Old Forge's Big Moose, among others, drew skiers before World War II stanched the flow of recreation seekers into the Adirondacks. Lake Placid, host of the 1932 Winter Olympics, which many gave credit for increasing the visibility and popularity of skiing in the United States, featured ski runs at North Notch, Scott's Cobble, Mount Whitney, and Fawn Ridge. Even owners of Lake Placid golf courses capitalized on the growing skiing craze by constructing lodges and tows on their gentle hills.[24]

Private businesses, sports clubs, and local governments ran the aforementioned centers, but beginning with the III Winter Games the state got in on the act of developing the Adirondacks for skiing. In preparation for the Olympics, the Conservation Department cleared nearly one hundred miles of ski trails, including a thirty-mile cross-country path around Whiteface and Franklin Falls Trail up the mountain's north side.[25] After the games, the Conservation Department built an expert ski trail parallel to the Mt. Van Hoevenberg Bobsled Run. In 1937, department staff began surveying the Adirondacks and Catskills for possible trail sites that would be "self-concealing" and cause "no defacement of the landscape or destruction of the wilderness character of the Forest Preserve." In other words, workers scanned the ground in search of existing truck trails, log roads, foot trails, and fire lanes that could be converted into paths for skiers without endangering the surrounding "forever wild" lands.[26] By 1939, the ski-trail construction program had yielded sixteen finished trails and three nearly completed, as well as seven in the planning stages.²⁷ With the scenic highway increasing its accessibility and attracting thousands, Whiteface Mountain became a prime target for development.

Notwithstanding the Conservation Department's trailblazing during the 1930s, the "forever wild" provision limited alteration of state lands for the benefit of skiing enthusiasts. According to John T. Gibbs, the Conservation Department's deputy commissioner, "The State can do very little to lighten the burden of New York skiers, who are going in increasing numbers each Winter to the more alluring 'snow spots' in New England because of the severe restrictions imposed by Article VII, Section 7." He added, "Narrower, continuous turn trails have been approved by the Attorney General as within the law, but the question of widening the paths like the other States is out of the question at present since it would interfere with the wilderness of the forest preserves."²⁸ Gibbs and other administrators worked inside a box created by state actors of decades past, who had imposed strict land-use restrictions on the Forest Preserve. Much like the assemblyman Fred Porter, who had seen his earlier efforts to build a bobsled run on state lands and pass the recreational amendment thwarted, in the mid-1930s state planners were bristling at the constitutional safeguards that prevented them from exploiting the full potential of the Adirondacks' mountainous terrain to provide for the enjoyment of downhillers and the economic benefit of the state.

Consequently, private groups took up the cause. In 1938, local skiing enthusiasts and businesspeople formed the Lake Placid Ski Council to promote development of Whiteface. The organization began with a drive to raise money for a one-and-a-half-mile downhill trail on a stretch of privately owned land situated on Whiteface's eastern slope. By year's end, the council had nearly completed the Whiteface Racing Trail and had dedicated the site in a public ceremony. The trail, however, was intended only for expert skiers and thus had limited significance in terms of mass recreation.²⁹ Further, lacking any infrastructure like tows and lifts, the trail was difficult to use. Lake Placid resident J. Vernon Lamb Jr. remembered skiing the trail as a teenager, when he and his friends used to say, "'Three hours up and three minutes down,' because we had to pack the trail out on the way up."[30] Not until technology had facilitated recreation seekers' engagement with the natural environment could skiers across all skill levels enjoy the mountainscape en masse. Still, the racing trail served as a means to a greater end: the construction of a ski center on Whiteface Mountain with modern facilities and trails for all downhillers. For Whiteface to be developed further, however, the state's legislature and voters would have to pass another constitutional amendment.

Since Whiteface Memorial Highway was open to motorists only from May through October (weather permitting), lawmakers, particularly local representatives who saw great economic potential in skiing, were receptive to plans that would enable vacationers to use the mountain during the winter. They wanted to see state conservationists continue to work closely with North Country businesspeople to promote tourism in the Adirondack Park. They got their wish in April 1940, when the legislature passed for the second time a resolution to amend article XIV, section 1 (until 1938 known as article VII, section 7), of the constitution, this time to permit construction of not more than twenty ski trails, between thirty and eighty feet in width, on the slopes of Whiteface.[31]

The Conservation Department endorsed the resolution's adoption as amendment 4. The department argued, first, that more skiing facilities would enable New York to compete with New England's skiing areas. This was a common argument among park administrators, businesspeople, and local and state lawmakers, who assumed that a state-run ski center in the Adirondacks would *compete* with New England's tourist spots while it would *complement* businesses in the North Country.[32] Second, in a telling bit of logic, the department asserted, "the development of such a center on Whiteface, which *lost its pure wilderness character* with the construction of the Whiteface Mountain Highway, will bring no small economic benefit to a large surrounding territory."[33] The Adirondack Mountain Club's skiing committee, the newly formed Whiteface Area Ski Council, and the editors of the *Plattsburgh Daily Press* all echoed the department's argument and insisted that Whiteface was no longer part of the wilderness.[34] According to this line of thought, existing development had nullified "forever wild," and past improvements should lead to more improvements. Once Whiteface had lost its "pure wilderness character" with completion of the road, it was no longer pristine and there was no turning back the tide of progress.

So powerful was the notion that once wilderness was lost it could not be recovered, so persuasive were skiing advocates, that even groups that had opposed earlier construction projects on Whiteface changed their tune when the skiing amendment came up for a popular vote. To be sure, several groups, including the New York branch of the Appalachian Mountain Club, the New York–New Jersey Trail Conference, and the Adirondack Wilderness Committee, opposed the "exploitation of public lands" and the "setting [of] an undesirable precedent for further future development of the park."[35] Nevertheless, the Association for the Protection of the Adirondacks, whose leaders had vehemently opposed the Olympic bobsled run and the Whiteface Memorial Highway, elevator, castle, and shelter house, expressed support for amendment 4. The association may have had a change of heart because skiing had a certain cachet that appealed to its prosperous members. Ski-center patrons tended to be in the middle and upper classes, for they had to pay for the necessary equipment, sportswear (men commonly skied wearing neckties), admission fees, lift-ticket charges, and other expenses. Moreover, since it was doubtful that anyone in the group was planning to build a Great Camp on Whiteface, their estates were safe from the possible ill effects of mass recreation there.

Whatever the impulses behind the association's shift in principles, the group began to embrace the rationales expounded by promoters of mass recreation. According to president Ottomar H. Van Norden, "Whiteface mountain already has a parkway to its summit, with a lookout on the very top and an elevator leading to it. *The mountain is no longer a wilderness area.* The proposed ski trails will give an added incentive to the use, in winter, of recreation facilities already provided by the state." By the early 1940s, the association was pragmatic about changing realities in the Adirondack Park and the growing popularity of outdoor activities like skiing. "As Conservationists," Van Norden added, "we must recognize that all the people of New York state own the state lands in the Adirondacks and we must not obstruct the proper recreational facilities that people demand and require from time to time, provided such facilities do not result in impairment of the wilderness aspect of our grand mountain park." In a reversal from its earlier antidevelopment positions that had earned the group a reputation as elitist, the association was now willing to sacrifice certain areas, particularly ones like Whiteface that had already been developed, in order to protect the park as a whole. Van Norden continued, "Only by recognizing this situation and by permitting the *reasonable recreational use* of the park in specified, limited areas, in which there is no real impairment of wilderness values, can its permanent wilderness character, as a whole, be preserved."[36]

Clearly, after the construction of Whiteface Memorial Highway, the definition of "reasonable recreational use" had changed to include activities that required much more elaborate infrastructure. As the wilderness inched closer to the city on the spectrum of environments, acceptable land-use practices changed. In fact, the association's contention that recreational development actually furthered "forever wild" echoed the argument of the assemblyman Fred Porter, the organization's foe in the earlier disputes over the Olympic bobsled and the recreational amendment. Soon after the association endorsed the project, on November 4, 1941, New York voters approved the amendment allowing for construction of "a skier's paradise" on Whiteface. The margin of victory was quite slim, however. It passed by less than 1 percent of the votes cast, and majorities in forty-nine of the state's sixty-two counties voted down the measure. The more populated parts of the state, where city dwellers yearned for contact with nature they did not enjoy at home, ensured the amendment's passage.[37]

In fall 1941, plans for the ski center were put in motion. The Conservation Department and the Whiteface Area Advisory Council, a private group employed by the state agency to offer advice on planning the new skiing area, chose to construct the Whiteface Ski Center on a site they believed to be accessible, conducive to optimal skiing conditions, and relatively inexpensive to develop. They selected a site on Marble Mountain, the twenty-seven-hundred-foot-tall northeast shoulder of Whiteface. Their rationale was twofold: first, Marble's slopes were smooth and less expensive to build on after loggers had cleared large swaths of formerly tree-covered ground when the land was in private hands. Second, and more important, situating the new ski center on Marble would take advantage of Whiteface Memorial Highway. The road twists from Wilmington through mountainous terrain on its way up Whiteface, and part of its path rests on Marble. Thanks to the highway, buses and Sno-Cats would be able to carry skiers up the mountain.[38] The plans, however, would not be implemented until after the war's end.

Yet even with the country embroiled in global conflict, lawmakers looked to the future and planned for an Adirondack Park that would welcome throngs of skiers. In April 1944, Governor Thomas E. Dewey signed the Whiteface Authority Act, which, in addition to authorizing construction, replaced the Whiteface Highway Commission with the Whiteface Mountain Authority to oversee both the highway and ski center.[39] According to Dewey, "The natural resources of this state are great enough and expansive enough to provide the kind of natural facilities for each kind of person without minimizing the enjoyment of any. Our great forest preserve and natural attractions will not be marred, except for this one mountain and this but slightly." Dewey acknowledged preservationists' criticism of the skiing development, recognizing that its construction would "slightly mar" the mountain, but focused on the recreational benefits for New Yorkers. According to the governor, compromising the scenic beauty of one among many attractive features in the Park was the price to pay for increased human access and enjoyment of nature. The *Lake Placid News* celebrated the bill's passage: "We have the location and terrain for a great recreational center. And now tied in with the Memorial highway and the Whiteface authority it will increase the popularity of that road in summer, after the war. The two will dovetail in attraction, the road in summer the ski runs in winter."[40] The fates of the highway and ski center were intertwined, but as we will see in the next chapter, the road's popularity did not ensure the skiing facility's success.

CONCLUSION

The United States' involvement in World War II delayed further development of the Adirondack Park as a playground, but only temporarily. By the time the country entered the war as a combatant, the state enjoyed a broad base of support behind its recreational-development programs. Vacationers from across New York State (and elsewhere) took advantage of the facilities that made camping, bobsledding, and mountain climbing more accessible. Even the Association for the Protection of the Adirondacks, the state's staunchest defender of "forever wild," came to accept development of Whiteface as a means of protecting the Forest Preserve as a whole. Although ordinary Adirondackers were ambivalent about tourism, a great many of them worked in the industry and themselves used outdoor facilities. In order to expand the tourism industry that employed and entertained so many, North Country lawmakers and business leaders worked in concert with the state.

However, discordant sounds of protest disrupted the superficial harmony achieved among competing players. As we saw in the previous chapter, during the early 1940s seasonal residents began to reject the cities of tents being built near their vacation homes. As some second-home owners pointed out, campsites often did not deliver the comfortable and healthful experience promised by Forest Preserve administrators. Moreover, the Whiteface Memorial Highway proposal sparked resistance among preservationists; and, even with the support of the Association for the Protection of the Adirondacks, the skiing amendment won only a slender majority of voters. As the state encountered new obstacles in achieving its goal of transforming the Adirondack Park into a modern wilderness playground after World War II, that fragile consensus would be torn asunder.

CHAPTER 4

A MOUNTAIN FOR ALL SEASONS?

New York State and Skiing on Whiteface Mountain, 1945–1971

In the April 14, 1967, column in her series "Sally's Skiing," Sally McMullen weighed in on the proposal for a fifth state-run ski center, this one planned for Hoffman Mountain in the Adirondacks' Blue Mountain Range. She wrote, "Pity the poor operators of private ski areas, struggling to compete against such odds! . . . Who in his right mind, would invest in a ski area, knowing that Big Brother might move in on the next mountain?" After New York State's twenty-year history of managing ski centers, McMullen was not alone in her skepticism about the doomed Hoffman amendment. From the 1920s through the 1940s, state conservation agencies functioned much like a chamber of commerce, promoting regional economic growth with the development of the Mt. Van Hoevenberg Bobsled Run, public campgrounds, and the Whiteface highway. With its substantial investment in skiing, however, Albany pinned many Adirondackers' hopes to an industry dependent on unreliable weather conditions. Although champions of recreational infrastructure on Whiteface were able to surmount legal roadblocks and secure constitutional amendments that allowed construction on "forever wild" lands, overcoming natural obstacles proved to be a far more daunting challenge. With the state eyeing another center on Hoffman, even after the mixed performance of its established skiing areas, Mullen concluded: "Welfare it is! Anyone for subsidized polo?"[1] As McMullen's comments suggested, the rift between private interests and the state was widening, and this chasm would shape environmental politics in New York's North Country for decades to come.

The first state skiing area opened at Marble Mountain in 1948, and, after the state abandoned the Marble site, the second began welcoming skiers to nearby Little Whiteface Mountain ten years later. According to state administrators and development advocates, a recreational landscape could not remain static; it had to adapt to changing notions of recreation and rising demand for outdoor fun. The growing popularity of skiing, however, posed challenges even more difficult to overcome than those encountered at public campgrounds. Campsite improvements were intended to create a more comfortable environment for vacationers—spaces where state-employed builders and caretakers, more than the campers themselves, performed the hard labor of maintaining a safe and healthful environment. Once defined as a space for mass recreation rather than as "forever wild," Whiteface, like public campgrounds, required constant maintenance and expansion to accommodate increased human pressure. The highway, and the subsequent projects it made possible, enabled recreation seekers to engage with the natural environment in new ways—mediated by technology as simple as skis and as mechanically complex as an elevator inside a mountain.

Nevertheless, improvements tended to blind planners to the mountainscape's imperfect potential to provide recreation. Even in an increasingly human-centered environment, natural changes significantly shaped Whiteface and people's experience of a mountainscape where rough terrain, powerful winds, and unpredictable weather made the site much harder to manage. The success or failure of recreational facilities hinged in large part on the degree to which they eased recreation seekers' contact with nature. If, as a result of poor planning, overwhelming human pressures, or uncontrollable natural phenomena, that contact became inconvenient, uncomfortable, unsafe, or unhealthful, the facilities required either substantial improvement or replacement. The quest for a middle ground that balanced the best features of wilderness and city on Whiteface followed an uphill path fraught with peril. Natural conditions threw into sharp relief government agencies' inadequate planning, and in the process strained the state's relationship with the businesspeople who would go on to become fierce opponents of new environmental policies.

The most vocal critics of the Whiteface ski centers were not "forever wild" advocates like the Association for the Protection of the Adirondacks, but rather North Country politicians and business leaders who had previously championed state-directed improvements to recreational infrastructure for their capacity to generate revenue. Whiteface Mountain's transformation served as a precedent and a model for future projects in the Park, where more mountain highways and state-run skiing sites would be proposed and built from the late 1940s through the 1960s. During the 1960s, however, local lawmakers and businesspeople began to decry what they considered to be the mismanagement of the state's debt-ridden, taxpayer-supported ski centers. It was the job of state agencies, these critics believed, to encourage the greatest possible influx of people and dollars into the Adirondacks, and the persistent economic plight of many Adirondackers was their evidence that the state was not doing enough. In a significant shift, prodevelopment groups began to characterize the state's investment in recreation as "welfare" and as competition for private industry. As state administrators struggled to establish a productive recreational environment on Whiteface after World War II, the close cooperation between state agencies and private promoters of development broke down.

THE ICY SLOPE TO RUIN: THE CAUTIONARY TALE OF MARBLE MOUNTAIN SKI CENTER

By fall 1945, four years after the state had laid out plans for Marble Mountain, construction was finally under way; but, in an ominous sign of things to come, the building process would not proceed as smoothly as planners hoped. In mid-October, workers stopped blazing trails when almost two feet of snow fell on Marble. Further complicating matters, the rising price of material and labor in the postwar economy prevented the Whiteface authority from purchasing the necessary supplies and further delayed construction. Though the agency hoped that work would be completed by the following autumn, construction halted again in summer 1946 as a result of the authority's inability to pay its bonds. Work would continue soon thereafter, inspiring curiosity in the local wildlife. In October 1947, a bulldozer reached Marble's summit to begin clearing a road on which materials for the T-bar lift would be transported. Lower down the mountain two bulldozer drivers noticed a deer watching them with what a local reporter described as "an intently critical eye." The machine's revving engine startled the animal, who took off in the direction of a crew chopping down trees with chainsaws. The sound of heavy equipment signaled to the mountainscape's four-legged natives that they would have to make way for many more people, structures, and machines in lands formerly designated "forever wild." Despite the noisy work that had aroused the deer's curiosity, planners missed their next expected completion date in December 1947.[2]

There was much work to be done to make Marble a skier's paradise. In July 1948, the superintendent of public works began accepting bids for construction of the center's lodge, water system, and sanitation facilities. In order to supply clean water to the lodge, workers built a dam on White Brook, which meanders past the site between Esther and Marble peaks, and sanitation facilities.[3] In the effort to transform Marble into a healthful human-centered environment, the state had to construct water-supply and sewage systems like the ones at public campsites. In the process, the development of Marble ensnared the surrounding area in reciprocal transformation. The ski center would function as a kind of city on a mountain, increasingly reliant on its hinterland, which served as a supplier of both labor and natural resources, as well as a receptacle for recreation seekers' waste.

In November 1947, as the state was investing more resources in wintertime recreation, New York voters approved a constitutional amendment authorizing the state to build two more ski centers, in addition to the forty-six private facilities already operating in the Adirondack Park, and the Marble site that was under construction. The amendment authorized the state to redevelop Gore Mountain, the destination of the first snow trains and formerly a popular site for skiing, which had fallen on hard times in the face of new competition in New York and New England. Also thanks to the amendment, in 1947 Belleayre Mountain in the Catskills would begin welcoming skiers to its new state-operated facility. Whereas the Whiteface authority managed the ski center on Marble, the supervisor of winter sports facilities, a new post created by the Conservation Department, was in charge of planning and administering the skiing areas at Gore and Belleayre.[4]

Finally, the Whiteface Ski Center at Marble Mountain was ready to welcome visitors in mid-October 1948, but all interested parties would have to wait for the snow to fall. Unfortunately for those who had invested time, labor, and money into the project, and for those who waited to ski down Marble's slopes, the clouds did not cooperate. Although the Marble site ostensibly opened in December 1948, skiing conditions remained unsuitable for wide use into the following year. Meanwhile, a committee of state officials and luminaries was still waiting to dedicate the center in February 1949, when the group called off the ceremony. Poor skiing conditions continued into March in what turned out to be a lost season in the center's inaugural year. Frank Elkins, a winter sports reporter for the *New York Times*, informed readers: "Once again, there is no sport whatever at the 2,700-foot level of Whiteface, where New York State has spent most of its money for development, while at 4,100 feet there is lots of cover, forty inches, and good skiing."[5] Already in its first few months of existence, observers began to question the wisdom of building the ski center on Marble.

The Whiteface authority adjusted as best it could to the unpredictability of the weather, but continued to confront significant problems managing the Marble site. Workers added two new rope tows to bring skiers forty-four hundred feet up to Lookout Mountain, which looms over Marble. Sno-Cats began carrying visitors from the ski lodge, located twenty-seven hundred feet up Marble, and up the Whiteface highway to heated shelters located on Lookout. At last, on December

30, 1949, Governor Thomas E. Dewey officially dedicated the Whiteface Ski Center at Marble Mountain. The site's troubles were far from over, however. On May 6, 1951, a fire destroyed the lodge, leaving behind only the chimney and foundation. Before the end of the year, a replacement building was under construction, and it would be ready for the 1952–1953 season.[6]

Even though skiing conditions improved at Marble during the winter of 1952–1953, interested parties sought new ways to expand the center. The skiing site offered winter play and the highway provided recreation during the warmer months. What if, the Lake Placid Chamber of Commerce wondered, a feature of the ski center could be used during the summer? Francis Sullivan, chairman of the chamber's Events and Sports Promotion Committee, noted that the drive up Whiteface, though easier than an unaided climb, was not without its hassles. Stalled and overheated cars, and drivers waiting until parking spaces became available near the mountaintop, resulted in traffic congestion that sullied motorists' experience. To remedy these problems, the chamber suggested, first, adding more parking spaces. Second, members proposed converting the ski center's T-bar lift into a chairlift so that sightseers would be able to take in the surrounding natural beauty from elevated seats during the summer. "We feel that the two proposals have direct relationship," Sullivan said. "Operation of scenic trips on the chair lift from the Ski Center would relieve congestion on the highway by channeling a percentage of the traffic to the Center. Such traffic congestion is relieved when tourists stop over at the North Pole on the mountain itself."[7] The "North Pole" Sullivan referred to was Santa Claus's Workshop, a tourist attraction featuring gingerbread houses and cartoonish architecture, constructed on a small patch of private land on Whiteface and opened in summer 1949.[8]

As more human-made features were added to Whiteface and its surrounding peaks, and as attractions became more elaborate and brought in more people, the words of those observers who had described the opening of Whiteface Memorial Highway in 1935 as the transformation of the mountain into urbanized nature seemed prophetic. State ski centers were part of an increasingly competitive commercial landscape.

Indeed, before long Adirondack Park administrators began lamenting "the existence of unsightly and unkept commercial properties" on the approach to the Whiteface highway, and proposed purchasing more lands surrounding the roadway to prevent the construction of additional eyesores along its path.[9] Within a few decades of the highway's completion, it suffered overfilled parking lots, traffic jams, and unattractive overdevelopment—problems more familiar to people motoring down Broadway than to those enjoying a scenic drive through what once, in the minds of many, qualified as wilderness. These challenges plagued both state and private developers who witnessed the consequences of a transformation that they had helped bring to fruition.

Because of the popularity of these attractions and regardless of the problems they created, officials and advocacy groups outside Whiteface's orbit began to yearn for their own large-scale recreational facilities that might serve as an antidote to the chronic unemployment and poverty suffered by so many Adirondackers.[10] Even though the Whiteface authority consistently lost money operating the ski center—between $15,000 and $20,000 annually during its first five years—reports of people flocking to skiing sites reverberated throughout the region, and local groups began asking, when is our turn? For instance, several organizations in the Central Adirondacks began a campaign in winter 1953 to build a state ski center on Bald Mountain, located a few miles outside Old Forge in Herkimer County. The Bald Mountain movement petered out, however, while the Marble Ski Center, which between 1949 and 1957 accumulated some $292,000 in debt, continued to provide an unpromising model.[11]

By 1961, Marble Mountain Ski Center ceased to exist, the victim of a fate sealed even before construction had begun. Planners had advocated for the Marble site with the hope that the ski center would capitalize on the Whiteface highway's popularity. Yet in their attempt to exploit existing infrastructure, planners had not adequately considered the human-created and natural conditions on the mountain. The human capacity for reshaping environments appears to have blinded them to the mountain's limited potential to provide recreation.

Marble lacked optimal skiing conditions. It quickly became clear that snow cover at twenty-seven hundred feet was frequently inadequate. In addition, Marble's slopes were steep, discouraging novice and intermediate skiers from enjoying many of the trails. Making matters worse, because logging companies had cleared the mountainside while the land had been in private hands, frigid winds swept across the site. The gusts blew away the softest powder, exposing the rock below and freezing the remaining snow. Though skiing can be a dangerous sport in the best of conditions, ski patrolman John Dreissigacker remembered how hazardous Marble could be on a weekend with heavy, wet snow: "we'd have five to ten bad injuries, most of them breaks above the ankle. We picked up a skier . . . but I couldn't tell which way to turn his leg, because his foot was on backwards." Skier Andy Flynn recalled the times he had competed in races on Marble during the early 1950s, and had felt "lucky to finish without a disaster since the entire trail was solid ice and my skis did not have metal edges." According to Douglas Wolfe, who during the 1980s and 1990s managed the Atmospheric Science Research Center which, in 1961, began occupying the ski lodge on Marble, "this area had already been logged, disturbed, and changed over the years." Wolfe then glanced toward a tall tree bending in the wind and observed, "the flag tree tells you that this is not the place to build a ski center. It's telling you there's a hellacious wind problem."[12]

The failure of the Marble ski center was a consequence of planners' vision of the hybrid natural and human-made environment, a vision narrowed by perceptual blinders that failed to give appropriate weight to ecological changes and weather conditions beyond their control. The construction of the Whiteface highway and attendant improvements fundamentally altered not only the physical landscape and uses of the mountain but also perceptions. Once developed for extensive human use, the mountain's function as a revenue-generating recreational site predominated in planners' minds. Improvements added in the name of progress begot more improvements, and so logic dictated that the first ski center should capitalize on the location of the mountain road. Yet determining the ski center's site based on the road's location ignored conditions on Marble itself, which had also been shaped by earlier extractive uses. Building on infrastructure with more infrastructure while disregarding natural conditions—snow cover, ice, and wind—ensured the end of the Marble ski center. The changing natural environment shattered the twin notions that development should inevitably lead to more development and that once wilderness was lost it could not be recovered.

Nonetheless, the commitment to recreational development on and around Whiteface would continue even in the wake of the Marble Mountain debacle. As Adirondack Mountain Authority member William Roden noted during a May 1961 meeting, "We should have people think of [Whiteface Mountain] as a recreation area. Along with this thinking, a family area with picnicking along the highway and picnic tables at the [Ski] Lodge would mean more hours at the Mountain and increased income at the restaurant, souvenir shops, museum, etc."[13] Roden and others thought of the mountain and its environs not as an untouched wilderness, but rather as a recreational space host to all the uses that label implied, including driving, sightseeing, skiing, dining, and shopping. Although natural changes ultimately forced the state to abandon the physical transformation of Marble, planners' perceptual conversion was complete. The lesson they took from the failure of Marble was not to avoid building a ski center whose success depended to a great degree on favorable weather, but rather to plan a new skiing site that would better exploit natural conditions. The imperative of economic growth would continue to drive their schemes, even as disappointing results failed to match expectations.

A "MOST CRUSHING REJECTION OF NATURE": DEVELOPING LITTLE WHITEFACE

Almost immediately upon taking office in 1955, Governor Averell Harriman, an avowed skiing enthusiast, publicly expressed his desire for a new ski center on Whiteface, one that would succeed where Marble had fallen short. In spring the following year, Harriman appointed the Joint Legislative Committee on Winter Tourists, chaired by the assemblyman Robert Main of Malone. The committee held hearings throughout the state to solicit proposals for winter-sports development and use that information to make recommendations to the legislature. For boosters of skiing, recreational development meant economic growth. For instance, the president of the Lake Placid Ski Club predicted that improvement of Mount McKenzie, located near Lake Placid and Saranac Lake, would generate $4 million in annual revenue. By early December, the committee had received bids for state aid in developing McKenzie, Gore Mountain (which had been neglected since the 1947 amendment that had authorized its redevelopment), selected mountains of the MacIntyre Range in the Adirondacks' High Peaks, and Whiteface. While the committee could not promise that all of these sites would be developed, its members expressed their determination to create a ski center that would meet the competitive requirements of the *Federation Internationale de Ski* and the Olympic Games, in the hope of attracting world-class competition to the state. The most promising site was Whiteface, whose development had already begun thanks to the constitutional amendment passed in 1927.[14]

The Committee on Winter Tourists heartily endorsed a new skiing area on Whiteface, urging the state to move "right away" and loan the Whiteface Authority $2.5 million for development of another subsidiary peak: Little Whiteface. Seeking to avoid a repeat of Marble Mountain, Main and the other committee members recommended construction of one chairlift from the twelve-hundred-foot level to the two-thousand-foot level, and a second from the latter point to thirty-six hundred feet, which was nine hundred feet higher than the tallest lift on Marble. On April 11, 1957, Harriman signed the bill to authorize construction of the (Little) Whiteface Ski Center, and the project moved forward rapidly. The following month, the Department of Public Works began considering bids for construction of an access road from Route 86 to Little Whiteface, and hearing from firms ready to build a 195-foot-long, 27-foot-wide concrete bridge across the Ausable River, and three parking areas for four hundred cars. Next, public works began accepting bids for the clearing of almost ten miles of ski trails.[15]

Unlike the first site, building the Little Whiteface Mountain Ski Center, despite the ambitious nature of the project and severe winter weather during the early days of construction, proceeded fairly smoothly. The two double-chairlifts, once completed, would carry skiers near the summit of Little Whiteface, from where downhillers would enjoy a choice of trails leading to the base twenty-four hundred vertical feet below. Little Whiteface, in fact, was the only ski center in the United States with a two-thousand-foot drop. At the base of the mountain, skiers would be able to relax at a two-story lodge, occupied by a shop, classroom, cafeteria, and lounge. In mid-January 1958, skiers began to use Little Whiteface's slopes, after nineteen inches of snow had blanketed the mountain. The following week, on January 25, Governor Harriman dedicated the new facilities at the ski center's opening ceremony. Opening day was not without its hiccups, however. During the lower chairlift's initial run, the motor broke down, stranding Governor Harriman and several other riders high in the air, with icy winds giving them the shivers.[16]

Despite this minor mishap, Little Whiteface opened to great enthusiasm and acclaim. After the new center began welcoming visitors in January 1958, 109 inches of snow fell over the next eighty-eight days, giving downhillers a fine surface on which to ski. The chairlifts alone brought $100,000 into the Whiteface authority's coffers, nearly double what administrators said they had expected. The conservation commissioner estimated that skiers spent $750,000 in the surrounding area during the season. As Little Whiteface offered another enticement for vacationers to visit Wilmington, Lake Placid, and other nearby towns until closing time on April 15, skiing enthusiasts filled local hotels, lodges, and inns, encouraging their proprietors to expand.[17] With the initial success of Little Whiteface, made possible by favorable natural conditions, the Marble ski center remained open only on weekends and holidays until 1961, when the lodge was converted into the Atmospheric Sciences Research Center.

Though Little Whiteface closed for the season in mid-April, one human-made feature on the mountain would not sit idle as outdoor temperatures rose. The two-mile-long double-chairlift, the longest in the eastern United States, enabled sightseers to ascend and descend the mountain while expending little energy. The chairlift extended the ski center's season, as the lift could be used by sightseers from May to mid-October and then by skiers in the colder months. During the summer of 1958, for instance, some fifty-thousand visitors rode the chairlifts.[18] Technology enabled Whiteface to become a mountain for all seasons.

Transcending the rhythms and realities of the natural environment was possible only to a limited extent, of course. Lacking the ability to exert control over the weather, the Whiteface Authority found itself under attack as the bearer of bad news whenever poor skiing conditions kept downhillers away. The close alliance the state had forged with local businesses and commercial groups subjected Adirondack Park managers to harsh criticism whenever hope outpaced results. Once the skiing season slowed down after the holidays, innkeepers around Whiteface blamed the messenger and bemoaned the ski reports given by the superintendent Arthur Draper. One Lake Placid businessman accused Draper of issuing "conservative" reports on snow conditions and "bending over backwards" to provide accurate descriptions. In the absence of favorable weather, local businesspeople who relied so heavily on Whiteface wanted the superintendent to tweak his descriptions of conditions on the mountain and to see more reports of "good" and "excellent" skiing, which existed when "the base [of snow] is adequate all over the area and the surface is covered with snow of a smooth, soft, and even texture."[19] If unable to control nature in fact, one could control it in fiction. A deceptive ski report, however, would not have hid the fact that unpredictable natural conditions laid bare the limitations of human planning and, as a consequence, strained the relationship between the state and development proponents.

Though unable to command the snow to fall, state administrators did continue to build on the mountain to increase both its capacity and its appeal. In summer 1959, after the center had been open for only two seasons, new improvements began. The Whiteface Authority constructed a new beginner skiing area with a sixteen-hundred-foot T-bar lift that was capable of transporting eight hundred skiers per hour. New construction bore fruit. During the ensuing third season of operation, Little Whiteface welcomed about fifteen hundred downhillers on weekends, sometimes between twenty-five hundred and thirty-five hundred on holidays. The parking areas could accommodate up to six hundred cars and buses, and during busy weekends it was common to find automobiles lined up for three-fourths of a mile along nearby Route 86.[20] Finding space for one's car at Whiteface began to resemble the search for parking on Broadway.

With increased human pressure on Little Whiteface, visitors were not only lined up along the road; they also had to queue up outside the bathrooms. Patron Noreen Grady wrote the conservation commissioner to complain about "the inadequate sanitary facilities and eating facilities at White Face Mountain Ski Area." She explained, "White Face Mountain is a favorite with the [ski] Club, but waiting in line for a half hour to an hour to use these facilities takes a lot of joy out of the sport of skiing." Draper responded on the commissioner's behalf: "over the holiday period we were swamped. During the summer we built a new parking area, but even this proved inadequate, and again cars were parked along the highway. . . . It is apparent that we must make further improvements. One of these improvements, now under consideration for development during the summer, is a shelter higher up the mountain. Provisions will be made [for] sanitary facilities and eating facilities."[21]

Warren Schlickenrieder, the district sanitary engineer, observed in February 1960 that not only were toilets too few but also the system that provided water and disposed of waste was faulty. According to the engineer's report, "The sewage chlorinator doses both the sewage effluent and the water supply, the former at the sewage contact tank, and the latter at the inlet to the water contact tank." This setup violated the state sanitary code, which forbade "any direct connection between a sewage disposal and a water supply system"; and so Schlickenrieder recommended that the connection between the sewage chlorinator and water supply tank be severed. Furthermore, he noted, the garbage storage area in the kitchen was not large enough to hold the amount of trash accumulated on busy weekends.[22] Just as rangers struggled to serve the growing number of campers in the Forest Preserve, the Whiteface Authority struggled to accommodate visitors to the ski center—in the process putting their health at risk. The same problems that plagued

Adirondack campsites tarnished the leisure experience for many skiers at well-traveled Little Whiteface. As a consequence, the state was now trapped between two increasingly frustrated constituencies: recreation seekers and area business leaders.

In an effort to improve the management of state-run winter sports facilities, in April 1960 Governor Nelson Rockefeller signed a bill to replace the Whiteface Authority with the Adirondack Mountain Authority, which had jurisdiction over Whiteface Memorial Highway, Whiteface Mountain Ski Center, and Mt. Van Hoevenberg Bobsled Run. The new authority would also be charged with redeveloping the neglected Gore Mountain. Even with these new responsibilities, Whiteface remained a priority for the fledgling Adirondack Mountain Authority. In spring 1960, the agency began planning construction of the auxiliary shelter Draper mentioned in his letter to Grady, in addition to more toilets, additional parking, and other improvements. As lawyer and conservationist Peter S. Paine Jr. observed, "if Whiteface is to keep on an even keel with the ski centers of New England, additional facilities must be installed. The Whiteface Ski Center must continue to expand in such a manner that it will never play second fiddle to any ski development in the Northeast."[23] Developers and planners could not rest on their laurels, especially when competing with other public and private ski centers. Little Whiteface, once developed, became ensnared in the never-ending competitive climb to financial success.

With so many people depending on the ski center, the Adirondack Mountain Authority looked ahead hopefully to the 1960–1961 season as a potential banner year for winter sports. Unfortunately, the snow did not fall. Whereas much to their delight the 1959–1960 season had seen substantial snowfall, the following winter required extra commitment on the part of the authority's workers to make Little Whiteface skiable. Administrator Dick Brenna and his staff engaged in what he called "Operation Squeezo": "We've been squeezing the possibilities out of every inch of snow since we opened in December [1960]. This is one of the worst winters we've had in years." Snowfall had totaled 199.8 inches in 1959–1960, but the following year amounted to a mere fifty-five inches. To compensate for the clouds' stinginess, Brenna directed his crews to traverse the mountainscape in Sno-Cats equipped with heavy rollers to pulverize ice on the ground, and with earthmoving equipment to carry snow from the woods and pack it on the ski slopes. Workers routinely labored fifteen-to-twenty-hour days in an attempt to get the mountain in shape for skiers.[24] Despite laborers' efforts, as a result of inadequate snow cover, Little Whiteface opened for only sixty-two days during the 1960–1961 season, compared to 123 days the year prior. During only twelve of those days was skiing "good," a dramatic drop-off from the fifty-three days of "good" skiing in 1959–1960. As a result, revenue plunged from $324,160 in 1959–1960 to $76,951 the following year, leaving the authority over $4 million in the red.[25] Overcoming nature's unpredictability became an increasingly frustrating and costly endeavor.

Although the success of the Little Whiteface ski center was, to a large extent, at the mercy of the elements, local business owners continued to criticize the Adirondack Mountain Authority for its management of the site. In summer 1961, Jack Wikoff, former professional skier and proprietor of the local Sun and Ski Motor Inn, released a letter detailing what he believed to be the authority's sins. Wikoff had wanted an even larger facility situated at a higher elevation, and pointed to Mount Killington and Sugarbush in Vermont as examples of ski centers whose development was outpacing Whiteface's. "Success lies," he asserted, "in the direction and expansion on the upper levels, a T-bar or other small lift in the big bowl immediately above the upper terminus of the lower lift, a gondola lift to the top of the ridge immediately below Wilmington Turn and trails to match, access roads to the upper trails presently existing for maintenance equipment, some needed improvement of the present upper trails under the direction of a competent authority."[26] Wikoff made no estimate of how much these improvements would cost, nor did he acknowledge that Vermont's skiing sites were privately owned. Still, he was not alone: local chambers of commerce and area ski clubs applauded Wikoff's statement and appealed to the state to improve the facilities at Little Whiteface. Soon after Wikoff released his letter, a group of local businesspeople formed the Whiteface Action Committee. In addition to lobbying Albany for further construction on Whiteface, the action committee resisted state-directed redevelopment of Gore Mountain on the grounds that Little Whiteface had not fulfilled its promise, and that the private skiing industry did not need more "arbitrary competition" from a new state-run facility.[27]

As managers of state ski centers groped for success, a shift in perspective began to take place among the agents of development. Whereas they once had seen the state as an ally in economic improvement, by the early 1960s they began to perceive public recreational facilities as unwelcome competition for private ones. The state, then, was caught in a lose-lose trap, doing *both* too little *and* too much in the eyes of local businesspeople. While state agencies that were tasked with mediating between people and the Adirondack environment once had to contend with preservationists' opposition to their large-scale recreational projects, they now faced the slings and arrows of the very development advocates who had championed campsite development, the Whiteface highway, and the ski centers.

In its endless quest to appease local businesspeople, improve skiing conditions, and transcend the unpredictability of weather patterns, for the 1961–1962 season the Adirondack Mountain Authority installed snowmaking equipment at Little Whiteface. Hence, "Operation Squeezo" reached a new phase. Despite a ballooning debt problem, the authority was committed to spending money in order to make money, and spent $98,000 to install the largest snowmakers available. The first machines covered about fifteen acres between the area's halfway station and the base of the mountain, where the beginners' slopes and ski school were located. Three and one-half miles of pipe carried water from a forty-thousand-gallon-capacity dam to twenty nozzles through which the snow was spread.[28] In the words of the *New York Times'* Michael Strauss, "The most crushing rejection of nature was the giant installation recently of artificial snow machines at the New York State ski center on Whiteface Mountain." He added, "The machines simply do, on command, what nature sometimes refuses to do—put enough moisture into cold air to turn it into snow."[29] Technology now enabled the Adirondack Mountain Authority to create, rather than wait for, good skiing conditions on the mountain. Thanks in part to the new snowmaking equipment, the center provided 105 days of skiing during the 1961–1962 season, almost double the year before, despite snowfall that amounted to the below average total of 72 inches.[30] Encouraged by estimates that during the 1963–1964 season human-made snow cover enabled the center to welcome two thousand visitors each weekend it otherwise would have been closed, in 1964 the authority laid additional pipes all the way up to the highest slopes.[31]

The authority came to rely more and more on human-made snow in subsequent years. According to a 1966 press release, "Whiteface has no natural snow but the wide 80-foot Valley Run trail from the mid-station lodge to the Base Lodge is gorged with acres of artificial snow. Though some other areas [besides Whiteface] are using snowmaking equipment . . . Whiteface has the largest total snow spread." In 1966, the ski center's snowmaking equipment, which now required more resources from the surrounding environment, began siphoning water from the Ausable River.[32] The press release's boastful tone reflected administrators' pride in the ingenuity and technology they had employed to overcome the unpredictability of nature. They distinguished between "natural" and "artificial" snow, the latter created by an elaborate sprinkler system that blanketed the mountain with white powder.

Although the Adirondack Mountain Authority may have considered snowmaking a novel development, it was consistent with state planners' perception of, and relationship with, Whiteface and its surrounding peaks since the late 1920s. When the natural environment did not conform to people's designs and demands, in many cases they had the power to make conditions more favorable to outdoor recreation. In order to get more people to ascend a mountain, they built a road on it. When that road did not reach the summit, they built an elevator to bridge the distance. In order to get skiers to the mountain, they built lodges, lifts, and tows to ease their trips up and down the peak. When those facilities proved inadequate, they expanded them. When snow did not fall from the clouds, they made snow. When the snowmaking machines did not make enough white powder, they built more equipment. The mountains, while they were protected as "forever wild," provided both opportunities and obstacles. Once amendments overrode constitutional safeguards and removed legal impediments to development, state planners, administrators, and workers found themselves in a never-ending (and expensive) struggle to maximize those opportunities and overcome those natural obstacles.

Thus, there was always more work to be done. The state-run Gore Mountain Ski Resort, which had become possible thanks to the amendment passed in 1947, finally opened in January 1964. Though a violent rainstorm spoiled the site's opening on January 25, the dedication of the Adirondack Mountain Authority's new skiing area created hope that this once popular site would be revitalized.[33] Whiteface, for its part, would also see additional improvements. Acceding to the demands of local businesspeople to develop the upper section, or "upper bowl," of Little Whiteface, in 1965 the authority began planning a new lift and three additional trails. The lower section of the skiing site would also receive a makeover with the addition of a new beginners' chairlift and an expanded ski lodge.[34] By the end of the 1960s, Little Whiteface, much expanded since its opening a decade earlier, featured six lifts, twenty-eight trails and slopes, two lodges, a ski school, two parking lots large enough for twelve hundred cars, and snowmaking machines along the full length of the first lift.[35] These facilities offered more than simple trails down which skiers could glide. According to Gerald Buyce, superintendent of the state-run Belleayre Mountain Ski Center in the Catskills, "No longer is the skier regarded as an individual who is supposed to rough it, putting up with poor accommodations, poor food, or inadequate facilities. He is regarded today as the customer in any highly competitive field would be.... With the number of areas available today, the skiing public does not have to tolerate inadequate facilities."[36]

By 1967 skiing in the United States was a $750 million industry that catered to 3.5 million people. There were more than nine hundred skiing areas throughout the country, most of them concentrated in New England and New York. The Empire State alone boasted one hundred and twenty centers.[37] Some believed there was room for more. In November 1967, voters considered a constitutional amendment that would allow construction of thirty miles of trails on Hoffman Mountain, located in Essex County, about midway between Whiteface and Gore.

Proponents, who included the Schroon-North Hudson Winter Sports Council and the Lake Placid Chamber of Commerce, anticipated brisk business at Hoffman.[38] According to a flier urging voters to approve the measure, "Development of Blue Ridge Ski Center will place us in the forefront of the country's annual 1-Billion Dollar Ski Industry. Adirondack region's economy and the entire state's economy . . . can be bolstered, expanded and kept moving."[39] The Conservation Department, Adirondack Mountain Authority, Governor Nelson Rockefeller, the region's state representatives, including the state senator Ronald Stafford, and dozens of skiing groups, supported the proposed ski center.[40]

The Hoffman amendment failed to pass.[41] In the light of the razor-thin margins by which the earlier ski amendments had passed, perhaps this was not surprising. William J. Fountain, supervisor of the town of Schroon, blamed the measure's failure on the apathy of state residents outside the region toward Adirondackers' economic hardship, as well as the state government's allegedly unenthusiastic effort to promote the site. "Our own attempts to help ourselves by promoting Hoffman Mountain resulted in a sad defeat largely because of the lack of official state support," he said. "We lose millions of dollars each year in the Eastern Adirondacks to the Ski interests of Vermont, and it seems to me inexcusable that the Empire State, endowed with such magnificent mountains as the Adirondacks, already largely in state ownership, must and has taken a back seat to a small state like Vermont."[42] Notwithstanding Fountain's bitterness toward the state, which in fact did champion the proposal, strong resistance outside the government did play a role in the amendment's defeat at the polls.

The Adirondack Mountain Club led opposition to the proposal. The club of conservationists and recreationists, founded in 1922, had been an enthusiastic supporter of the two Whiteface ski centers. In the case of Hoffman, however, members asserted that New Yorkers would "lose the money for capital investment to build the development; they lose the Forest Preserve land it is built on; they are likely to lose more money to meet operating deficits—all to provide a 'loss leader' to attract business for local commercial interest." The club also feared a repeat of the Marble Mountain debacle, and claimed that snow depth at Hoffman would not sustain profitable business.⁴³ This conservation group expressed its opposition with tried-and true environmental arguments, but also borrowed the commercial rationales that businesspeople typically employed. The organization did not distinguish between what was good for the environment and what was good for people, as its members asserted that a ski center on Hoffman would both mar Adirondack scenery and cost New York taxpayers considerable sums of money. In light of Marble's failure and Little Whiteface's struggles, recreational developments considered reasonable during the 1940s and 1950s now seemed much less promising to many.

The Adirondack Mountain Authority's troubles would end in late 1968, however, when Albany dissolved the agency. The legislature voted to pay off the authority's $13 million in outstanding obligations and transferred jurisdiction over the Whiteface highway, Mt. Van Hoevenberg Bobsled Run, and Gore and Whiteface ski centers, to the Conservation Department. The department maintained its management authority over the entire Forest Preserve and the Belleayre ski center in the Catskills until it was replaced by the Department of Environmental Conservation (DEC) in 1970.⁴⁴ The conservation agencies also took on the burden of overseeing a new Adirondack project in June 1969, when the far-reaching impact of the transformation of Whiteface Mountain since the late 1920s became more evident with the opening of Prospect Mountain Veterans Memorial Highway. Whiteface's development had set a new standard in the Adirondack Park, one that other locales attempted to emulate. There was the push for more ski centers, which bore fruit with Belleayre and Gore, and there was also a movement for a road up Prospect Mountain, a twenty-one-thousand-foot-high peak in Lake George. Whereas visitors once had to endure an hours-long climb up Prospect, with the road open they now could drive up in ten minutes, leave their cars in an eight-hundred-car parking lot, and hop on a jeep-pulled trailer that carried travelers the rest of the way to the summit.⁴⁵ Now Whiteface was not the only mountain in the Adirondacks with its own Broadway.

Despite significant investment in the park's recreational infrastructure, the promised economic benefits to the region did not lift

Adirondackers out of poverty. In 1971, as the post-World War II boom was coming to an end nationwide, Essex County had an unemployment rate of 12.4 percent, almost double the national average. One out of eight families in the county earned an annual income below $2,000. According to James DeZalia, chairman of the county's Board of Supervisors, "Tourism is our biggest dollar producer. But the season only lasts from June to Labor Day." It was revealing that he ignored the less lucrative winter season in his calculations. As a *New York Times* reporter described the county's plight, "Lake Placid . . . is the goal of hundreds of thousands of tourists who each year frequent its luxury hotels and dine in the old-American elegance of the Whiteface Inn or the Steak and Stinger restaurant. But most local people here know the luxury hotels and restaurants only as waiters, waitresses, and kitchen help. Lacking the protection of unions and competing in a labor market flooded with college students during the summer, most employees in the tourist industry can command only the minimum wage."[46]

Skiing may have been a $750-million industry in the United States—with one hundred and twenty ski centers in New York State, many of them located on public and private lands in the Adirondack Park—but little of that money trickled down to the seasonal service workers on whom tourists relied. With state agencies struggling to manage Adirondack skiing areas, downhillers complaining about inadequate facilities, local businesspeople griping about tourism's disappointing economic impact, and many ordinary Adirondackers still struggling to eke out a living, the hope for a winter-sports renaissance in the region seemed to be buried in ice alongside Marble Mountain's unused trails.

CONCLUSION

Whiteface Mountain, a popular feature of the modern wilderness playground, since the opening of the highway had become more accessible over time. But once state agencies struggled in their efforts to merge human-constructed facilities with nonhuman nature in a way that best exploited the recreational potential of both, skiing at the state-run center increasingly became a privilege of those who could afford to pay for the rising cost of outdoor play. Skiing at Whiteface was becoming less possible for working- and even middle-class recreation seekers, particularly those Adirondackers whose meager earnings from working in the tourism industry prevented them from enjoying the very services their labor supported. Despite frequent promises of riches flowing into Adirondack communities, the region's sparkling, modern facilities masked for tourists the larger environmental, economic, and social problems that the recreational development that prevailed in the Adirondacks from the 1920s into the 1960s failed to address.

During the debates over post-highway construction projects, many came to the conclusion that building the road had deprived Whiteface of its distinction as wilderness. As a result, the mountain now became useful as a site for mass recreation. Subsequent developments continued the process of pushing Whiteface farther away from its past status as a wilderness area toward a more modern place featuring many of the comforts of home. Regardless of the new recreational features' profitability—indeed, neither ski center was self-supporting as intended—politicians, planners, and boosters continued to trumpet

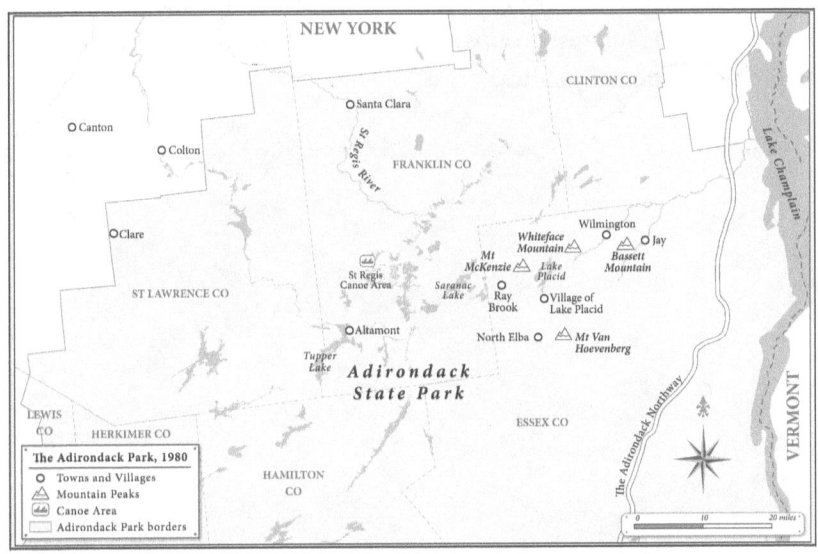

FIGURE 9. The Adirondack Park, 1980. Map by Gerry Krieg of Krieg Mapping.

the economic miracles that development would bring. Once a path was laid, as with the construction of Whiteface Memorial Highway, a new logic determined future alterations and functions of the landscape. Recreational uses that required elaborate infrastructure, even on an intractable mountainscape, predominated in the minds of development advocates, despite their failure to reap adequate financial rewards for locals and the state.

In recreational landscapes like the one built on and around Whiteface, natural features provided the opportunity for sightseeing and skiing, but it was up to state planners to exploit nature's potential. The natural environment, however, also proved to be an obstacle to recreational uses, because once those uses were defined, administrators perceived the landscape almost exclusively in those terms. Thus, exploiting nature's potential required taming it, protecting recreation seekers from the more harmful aspects of the environment, and also stepping in to do nature's work whenever it did not follow planners' whims. Yet as Albany committed more money into an undependable winter-sports program, the prodevelopment accord of the 1920s–1950s began to fall apart. As we will see in the next chapter, another road, the Adirondack Northway, would further splinter the shrinking consensus into competing interest groups and set the stage for a far more combustible environmental politics.

CHAPTER 5

ADIRONDACK SPRAWL

From the Northway to the Creation of the Adirondack Park Agency, 1959-1972

During the summer of 1969, the Temporary Study Commission on the Future of the Adirondacks, a state body created the previous year by Governor Nelson Rockefeller, held hearings throughout New York State to solicit the public's input on the fate of the Adirondack Park. At an August hearing in Lake George, Margaret Lamy, a native of Saranac Lake, a former editor of the *Lake Placid News*, and the director of publicity for the Adirondack Museum, described how her experience in the Forest Preserve had changed over time: "I use the woods and trails and waterways as much as anyone, and within the last five years I have become increasingly alarmed by what I consider to be the destruction of the wild character of this region." Her recent experience in nature failed to match the enjoyment of her earlier excursions, when open spaces appeared vaster and the forest seemed wilder. Lamy felt compelled to ask, "We have recognized the necessity in this country for planning for our urban areas. Why can we not see the need for just as careful planning for our wilderness areas?"[1] At a time when sprawling Forest Preserve campgrounds were the norm and massive vacation communities became a real possibility on the Adirondack Park's private lands, Lamy was not the only one asking this question. Once the Temporary Study Commission began looking for answers, the challenge of controlling growth in the Adirondacks was made more difficult by the state's inability to appease a restive population heavily invested in the Park's future.

As a vocal champion of stricter land-use controls in the Adirondack Park, Lamy was the exception among the region's residents, who tended to oppose the expansion of state authority. Her allies were mostly, but not exclusively, nonresident recreation seekers, second-home owners, and environmental activists. What they all shared was an appreciation of the park as a space for outdoor play. The expansion of recreational facilities and second-home ownership from the 1920s through the 1960s had given New Yorkers from across the state a stake in the Adirondack Park's future, and they banded together to preserve its wild character. The modern wilderness playground attracted people across class lines, and, as a consequence, a broader, largely middle-class movement took over for the blue-blooded Association for the Protection of the Adirondacks as the leading defender of Adirondack wilderness. Ironically, the facilities and dwellings that had enabled them to enjoy both state and private parklands were now considered the problem, for their continued expansion would enable new and larger groups of people to become campers and second-home owners. Just as locals embraced the Adirondacks as their own and defended it against invaders, so did recreation seekers and seasonal residents develop a sense of ownership over their piece of the park. Thus, at the heart of the region's contentious wilderness politics lay the question: *whose* park was it?

This question became much more difficult to answer after World War II, when recreational use of the Adirondacks exploded. Particularly after the Northway, the extension of Interstate 87 from Albany to Montreal, was completed in 1967, people flocked to New York's North Country in unprecedented numbers. What could be wrong with that? After all, state conservationists had, since the 1920s, encouraged mass recreation in the Adirondack Park because, they argued, it belonged to all New Yorkers. That democratic notion did not always lead to coherent policy, however. While administrators attempted to find a happy medium between recreational development and resource conservation, they were caught between champions of wild forestlands and Adirondackers who rejected the state's expanding role in their communities. By the early 1970s, the consensus surrounding earlier recreational developments would seem like a distant memory.

In a controversial attempt to remake park policy, in 1971 the legislature and Governor Rockefeller followed the Temporary Study Commission's advice and empowered the new Adirondack Park Agency (APA) to devise comprehensive zoning plans for both state and private lands. Yet even in the context of the blossoming environmental movement and governments nationwide enacting new development restrictions, changing course would not be easy.

WHICH WAY THE NORTHWAY? THE DEBATE OVER A NEW ADIRONDACK HIGHWAY

The development of the Adirondack Park during the interwar years had made the Forest Preserve legible to city dwellers and had laid the foundation for further improvements after World War II, when a booming economy, the continued proliferation of the automobile, and increased road building unleashed millions of new recreation seekers nationwide. Yet New York, which had been a pioneer in resource management and planning for outdoor recreation, and ranked second only to Maine in providing 170.5 acres of state parklands per one thousand citizens, would face tremendous challenges in attempting to accommodate a growing number of vacationers in its public lands. Even with an ambitious conservation and recreational-development program in place, the Conservation Department discovered that demand for play space far exceeded supply.[2] Attendance at the seventy-eight parks under the aegis of the State Council of Parks jumped to a record 20,811,878 in 1949. Ten years later, the number of visitors to the council's parks climbed to 31,232,263—almost double New York State's population in 1959. Use of ski centers, camps, and trails in the Catskills and Adirondacks grew substantially, as well: between 1924 and 1954, the state estimated that the total number of visitors to both parks increased from 23,000 to 1,577,000.³

The post–World War II period also saw a great increase in the number of seasonal dwellings in the Adirondacks. Owing to the postwar expansion of the middle class and the attendant rise of car and homeownership, the 1960 U.S. Census counted 39,000 vacation homes in the twelve Adirondack counties—a 75 percent increase over 1950.⁴ The Vanderbilts and Rockefellers, whose Great Camps had signaled the creation of a genteel pleasuring ground in the Adirondacks during the late nineteenth and early twentieth centuries, were giving way to the "Joneses": upper-middle-class strivers looking for their own piece of the Adirondack Park. As second-home ownership expanded in the 1950s, more Americans endeavored to slow further growth and protect the park's natural environment. The proposed Adirondack Northway had the potential to bring many more recreation seekers and second-home owners to the North Country.

The Northway provoked controversy even before it was completed, however, because the proposed route cut through 254 acres of "forever wild" land. Since the Department of Public Works selected "route B" through the Forest Preserve, the highway, like the Whiteface road and ski centers before it, required a constitutional amendment to become a reality. Anticipating a fierce public debate, public works consulted with the Conservation Department in an attempt to allay the fears of "forever wild" purists. Planners promised that a mere 0.02 percent of the 2.2 million acres of state forestland would be affected: "The scant preserve acreage Route 'B' will traverse is for the most part in scattered locations—along small 'islands' on the eastern fringe of the preserve. . . . These separate parcels of the preserve area are already penetrated by State and local roads. They therefore are not and never can be a real wilderness."⁵ Development advocates continued to place environments on a spectrum between untouched wilderness and built-up city as a way to justify new projects, and the areas where construction would take place were not "real wilderness" and never could be because of the presence of human-built roads. The absolute definition of a pristine wilderness thus served to undercut preservationists' goals. If wilderness needed to be pure, any human incursion could be evidence of a paradise lost and never to be recovered. Consequently, the established pattern in which existing improvements led to more improvements could persist unabated.

The Northway, like earlier state projects in the Adirondack Park, was supposed to be another mechanism for the democratization, recreational development, and financial enrichment of the region.[6] The Department of Public Works went so far as to assert that, in addition to increasing access, the highway would also become an integral part of the scenic and recreational environment: "Route 'B,' like a modern parkway, will be so constructed as to blend with the landscape." After all, if electrified campsites, mountain roads, and ski centers could become an accepted part of the Forest Preserve, then why not the Northway? Public works promised that restriction of commercial and industrial development along the road would "keep intact the scenic beauty of the Adirondack area." The highway would also benefit hunters, hikers, anglers, and even wildlife: "The new superhighway will neither block streams, fill in ravines, nor bar passage between sections of the preserve. Eighty-eight Northway structures (bridges) . . . will permit the movement of wildlife and hikers through all segments of the preserve. . . . The Department [will] provide walk ways at stream crossings to accommodate hunters and fishermen." These proposals to ease the movement of recreation seekers and wildlife won the support of the Council of Conservationists, the Adirondack Park Association, and other conservation groups that embraced a utilitarian vision of the Adirondacks. Hoping for jobs and commerce, construction unions and upstate farmers also lobbied for the Northway. Their efforts were successful when the state legislature adopted the route B resolution for a second time in 1959 and sent the amendment on its way to a popular vote.[7]

Lawmakers and planners hoped the Northway would reinforce mass recreation in the Adirondack Park. As one of the managers of the recently opened Little Whiteface ski center, Arthur Draper was enthusiastic about the highway's potential to boost the popularity of the region's state-run facilities. In a letter to the *New York Times*, Draper recalled how thirty years earlier opposition against the Whiteface highway had surfaced, and yet "Each year thousands . . . stand on the summit and marvel at the gorgeous beauty and mystery of forested mountains and lakes." He added, "To say that the proposed Northway will mean the end of the forest wilderness is hyperbole. It will mean rather that thousands more will be brought into the Forest Preserve and that many of these thousands will have an introduction to what conservation really means—*wise use of natural resources*—and a love for a sacred heritage."[8]

Once again, development of the Adirondack Park fit within a particular conception of "wise" conservation—one emphasizing increased access to scenic nature and efficient use of resources over strict wilderness preservation. Significantly, three decades later conservatives would adopt the slogan "wise use" to champion private property rights and development at the expense of environmental protections.[9] As we will soon see, the gap between environmental activists and prodevelopment groups would become a yawning chasm well before the antiregulation, "wise-use" movement organized during the late 1980s. Despite Draper's dismissal of preservationists' concerns, however, there was much evidence to demonstrate that development begot more development. If Draper had fleshed out his narrative, the amendment's opponents would have found additional arrows to add to their quivers, including the failure of the Marble Mountain Ski Center and the destructive effects of mass recreation at campgrounds. Indeed, once the Northway was completed and its far-reaching impact became evident, many would question how "wise" established land-use practices were.

Moreover, Draper was arguing against a straw man, because the Northway conflict, like the Olympic bobsled controversy almost thirty years earlier, was over *where* the facility should be located rather than whether it should be built at all. Some "forever wild" purists favored a route that would not cut through the Forest Preserve but instead traverse private lands in the Champlain Valley, east of route B.[10] A group of mostly middle-class nature enthusiasts met in Schenectady in December 1958 and formed the Citizens' Northway Committee to spread the word in favor of the alternative route C. The committee warned New Yorkers, "Once you let industry in—the wilderness is doomed. . . . Man can always build a road, but he can never rebuild a wilderness."[11] Since most Northway committee members lived outside the Adirondack Park and knew it only as visiting vacationers, they wanted to maintain the integrity of the Forest Preserve, based on decisions made by state actors and landowners since the 1880s that had set aside Adirondack forestlands as sites for leisure-time fun.

The Northway committee was not anti-development, however. According to the committee, the Champlain Valley, "already partially industrialized, badly needs to link up with the Interstate System." The group was not challenging the arguments of route B proponents based on the necessity of a highway through the North Country. Rather, its members were willing to sacrifice a different landscape not protected by article XIV, section 1, of the state constitution. In fact, the committee made the case that the Champlain Valley route would spur industry in an already partially developed area that boasted "three times the year-round population and about six times the factory employment."[12] Though a number of environmental groups joined the citizens' committee as route C advocates, including the Adirondack Mountain Club, Sierra Club, and Wilderness Society, they were also joined by a few prodevelopment organizations.[13] The Warrensburg and Essex County chambers of commerce, and the U.S. Route 9 Improvement Association, agreed with the committee that the Champlain route would provide a greater financial boost to the region.[14] Yet despite the efforts of the Northway committee and its allies, on November 3, 1959, New York's voters approved route B.[15] Still, the fighting was not over, for the Northway debate served as a rehearsal for more heated conflicts over land use during the late 1960s and 1970s.

As lawmakers and residents considered new modernizing improvements like the Northway, the pull of the past and the pull of the future were strong. For champions of "forever wild," it was not the attraction of an unchanging wilderness but rather the tug of past legislation and land-use practices that had established a recreation-centered landscape in one spot but not in another. They were willing to see private lands paved over to encourage industry and commerce—but not state lands. For some prodevelopment groups, the prospect of a more prosperous future moved them to champion route C. Although in the Northway debate the interests of nature enthusiasts and some development proponents dovetailed, this fragile accord based on a shared willingness to develop the park's private lands would shatter once the impact of the Northway became evident. As the highway increased access to the park, politicians, businesspeople, and developers sought new ways to entice more vacationers and seasonal residents into visiting the Adirondack Park. Though only a few dozen miles between routes B and C separated the two sides in the Northway debate, the gap between environmental activists and development proponents would become an unbridgeable, dragon-infested moat once sprawl began to gobble up both public and private lands.

CAMPGROUND SPRAWL: COPING WITH SUCCESS AT STATE CAMPSITES

The tremendous postwar increase in the number of recreation seekers, encouraged in part by the Northway, was a significant development and a substantial challenge for state planners, but equally significant and challenging were the kinds of recreational activities New Yorkers and Americans at large were pursuing. A survey conducted by the Conservation Department in 1960 identified several major trends in recreation. The department observed that outdoor play was booming as a result of general economic prosperity and increased mobility made possible by new transportation options like the station wagon. Families, the survey results revealed, were more often vacationing as a unit, and they were interested in what the writers of the report referred to as "participant sports, particularly those which take place outdoors." More specifically, they later added, "Camping is the fastest growing form of outdoor recreation in New York."[16] This was true nationwide as well, and the federal government responded with the National Park Service's "Mission 66" and the Forest Service's "Operation Outdoors" in the effort to expand camping facilities in national parks and forests.[17] In New York, as elsewhere, vacationers took full advantage of the facilities for active recreation that state administrators had begun building during the 1920s, and planners more than ever struggled to keep pace.

The Conservation Department's 1960 report emphasized the need to improve and maintain existing facilities better, and to acquire new sites for campground development. The department recommended that the state should develop, between 1960 and 1965, twenty-three new campsites in addition to the thirty-eight already located in the Forest Preserve, and an additional forty-nine in the following decade. Whereas concentration had been the rule prior to World War II, suburban-style sprawl came to define campsite construction after the war's end. Between 1920 and 1950 the state had built thirty-eight large campsites in both the Adirondacks and Catskills; in 1960 planners recommended almost doubling that total with seventy-two new camps over the ensuing fifteen years. Campsites could not be built anywhere, however. According to the department, "An attractive setting for tents and trailers, sanitary and garbage disposal facilities, clean water to drink and swim in, and the supervision of a campsite caretaker—all this is what our own people and guests . . . expect to find at our public campsites." In attempting to provide the comforts of home at state campgrounds, administrators catered to a specific kind of camper with particular expectations. "Public campsites are not patronized to any substantial degree by the rugged type of camper," the Conservation Department observed. "They are used by the family group who have a liking for the out-of-doors, but *who need some degree of comfort and convenience*, and by those who must, out of economic necessity, seek an inexpensive vacation or have no vacation at all."[18]

The "public" that these campsites served comprised urban and suburban families who wanted to engage in physical activity and enjoy the novelty of being in the outdoors, but without the trials associated with the wilderness experience. By providing elaborate infrastructure to increase the comfort level of families unaccustomed to life in the wild, the Conservation Department attempted to avoid creating a new generation of "Murray's Fools." In addition, improved facilities were intended to prevent the breakdowns in trash collection, water supply, and waste disposal that had plagued campsites during the interwar years. If left to their own devices—without roads; without functioning water and sanitation facilities; without electric lights, stoves, tables, and benches; and without a caretaker's supervision—campers' enjoyment of the Forest Preserve might be spoiled by the toil, anxiety, and discomfort that came when one lived without the conveniences of home.[19]

Stemming from the Conservation Department's characterization of campers as families seeking some measure of comfort in the outdoors—the kind of comfort that only human planning and labor could provide—administrators invested more resources in further domesticating public campsites. During the early 1960s, the engineering unit within the department's Bureau of Forest Recreation assumed greater responsibility in campsite construction. According to the Temporary Study Commission on the Future of the Adirondacks, the ascendancy of the engineering unit "has led to the construction of highly sophisticated campsites with sewage systems and more concern for the facility than for the land it is being constructed on. In addition, larger volumes of timber have been removed than in the past." As a result, the commission reported, the "forever wild" clause of the state constitution "is not being strained by the existence of campsites, but by the changing character of campsite construction." The study described the campsites created from the 1920s through the 1950s as "primitive," but this would have been news to early administrators. As we have seen, the Conservation Department had gone to great lengths to improve campsites with facilities for providing water, disposing of waste, and illuminating the grounds.[20] However, the growing scale of recreation necessitated, in the state's view, more sophisticated and technologically advanced means of accommodating millions more campers accustomed to a greater degree of physical comfort in their day-to-day lives.

The Conservation Department's plans for improving the Adirondack Park's most popular campsites, paid for by a $200 million bond issue approved by New York voters in November 1966, revealed the extent to which administrators would go to improve public campsites. For example, the ever-popular campgrounds at Fish Creek Pond were due for another makeover in the wake of the post–World War II recreation boom. What began as a single campsite with a few fireplaces during the early 1920s, featured in the mid-1960s 382 individual campsites with the capacity to host 1,490 overnight campers and 900 picnickers. For comparison, of the 107 towns located either entirely or partially within the Adirondack Park at the time, thirty-nine had a population of less than 1,000.[21] If strength came in numbers, the growing population of vacationers acquired even greater clout in shaping the Forest Preserve, and they would exert their collective power in the years ahead.

At Fish Creek Pond, the means by which one measured quality of life at home could also be used to gauge comfort and convenience while engaging in outdoor recreation. The campgrounds there had three different water supplies: one originally installed by the Civilian Conservation Corps during the 1930s, the second in 1963, and the third in 1966. The Conservation Department proposed, in 1966, that the oldest water supply system be abandoned and replaced by a "new hydropneumatic system . . . similar to No. 2 and No. 3." The second and third systems, which planners hoped to emulate with a new pump, featured pumps powered by an electric motor, with 2,000–2,500-gallon pressure tanks and air compressors. In addition, the twenty-four bathrooms for camp visitors were deemed "undersized" and in need of more toilets and septic tanks. The Conservation Department further observed that the electrical facilities, provided by Niagara Mohawk Power Corporation, were "in poor condition" and in need of replacement.[22] The quality and size of the water, sanitation, and electrical facilities at Adirondack campsites determined their capacity, and improved amenities would not only increase comfort for campers, but also enable individual sites to welcome more visitors. Despite these measures, however, the problem of overloading continued. During the 1968 season, for instance, park rangers had to turn away 6,736 campers from Fish Creek Pond.[23] The Conservation Department recommended similar improvements to seventeen more Adirondack campsites, as the effort to domesticate public campgrounds for a growing group of vacationers continued in earnest.

In the context of the automobile-borne recreation seekers from cities and suburbs flocking to campsites that featured many of the conveniences they enjoyed at home, one may be tempted to ask: what happened to "forever wild"? As the Adirondack Park continued to develop and the post–World War II environmental movement grew in size and influence, many in the region and throughout New York State asked this very question. Substantial changes made to the Adirondack Park's state lands since the 1920s rarely occurred without a fight; but since campsite improvements occurred incrementally, often built on existing infrastructure, and catered to a large number of people, they went relatively uncontested. That was about to change.

Since the First World War, state-directed recreational development had given a great many New Yorkers a stake in the Adirondack Park. During the 1920s, state conservationists began to forge a middle ground that combined the comforts of home with the novelty of wild nature, and in the process they created a much broader clientele for state parks. Those development proponents who had championed the Lake Placid Olympics, campsite improvement, Whiteface highway, and state-run ski centers on the grounds that new facilities would create new nature lovers were absolutely right. However, they could not have predicted how that appreciation would be cultivated and expressed. The initiative to promote outdoor play worked quite well: roads gave hundreds of thousands of people access to green spaces, and the amenities provided by the state enabled them to enjoy nature without concerning themselves with much of the dirty work involved in the outdoor experience. The hassles they encountered (polluted campsites, malfunctioning toilets, tainted water) appeared to be the result of human failings (managers' inadequate planning or other campers' carelessness) rather than the exigencies of the natural environment.

By the early 1930s, a growing number of campers and hikers began to venture beyond crowded campsites to feed their hunger for a purer form of wilderness; and early in the next decade, second-home owners began to protest the public health threats posed by polluted campgrounds near their properties. After World War II, both short-term recreational uses and second-home ownership grew dramatically, which created more nature lovers and simultaneously seemed to threaten the scenic, rugged, healthful nature they had come to cherish through recreation. Once they began to perceive increased access and infrastructure improvement as antithetical to wilderness, they petitioned the state to halt development and to institute stronger safeguards for Adirondack nature. As these mostly middle-class nature enthusiasts took up the cause of "forever wild," they replaced the blue bloods of the Association for the Protection of the Adirondacks as the leading defenders of the park.

As a consequence, the broad consensus on campsite development would crumble in the late 1960s, especially as New York's nature enthusiasts joined with national environmental groups in an effort to shape land-use policy in the Adirondacks. During the interwar years, New York groups, most notably the Association for the Protection of the Adirondacks, had led the fight to save "forever wild." After World War II, the ecological stakes seemed higher nationwide as the threats that inspired the environmental movement, including nuclear fallout, chemical pesticides, and toxic petroleum-based products, appeared cataclysmic. Among the greatest concerns for environmentalists were overdevelopment and the consequent loss of open space. As residential and commercial development sprawled outward from American cities with the aid of parkway and highway construction, reformers from coast to coast assailed the disappearance of the green spaces they treasured as wildlife habitats and recreational spots.[24] The Sierra Club and Wilderness Society were involved in the 1959 Northway debate, but they took a backseat to the Schenectady-based Citizens' Northway Committee. National groups would go on to play a more prominent role in the ensuing years once postwar developments further highlighted the contradictions inherent in maintaining a park that included both public and private lands. Could the transformation of the Forest Preserve spell the end of the recreation-centered Adirondack Park? By serving the majority of campers and picnickers who were visiting the Adirondacks in order to enjoy an environment more amenable to significant human presence, were administrators ignoring wilderness lovers?

In the aforementioned 1960 recreation survey, the Conservation Department did acknowledge the growing popularity of what it called "a national 'back to the woods' movement of major proportions." This movement was indeed on the rise, but it was not new. As we have seen, in 1932 Adirondack Park rangers began counting campers and hikers who trekked beyond domesticated campsites to more rugged parts of the Adirondack wilderness. In response to the escalation of this trend in the postwar years, the department identified eleven separate tracts totaling 664,200 acres of Forest Preserve land (out of a total 2.5 million acres) within the Adirondack Park that were large and remote enough to be considered "real *wilderness* areas." The report recommended that the state consolidate these holdings, acquire private properties within them, and provide hikers with marked trails and lean-tos.[25] Even in areas described as "real wilderness," the state endeavored to make them legible to both administrators and visitors.

The state had attempted since the 1920s to balance competing interests in the Adirondacks, but this effort was complicated by the park's puzzle of state and private lands that catered to a wide variety of stakeholders. As the scale of recreational development and uses grew after World War II, the state seemed caught between, on one side, a large number of persistent development advocates and, on the other, increasingly vocal groups of nature enthusiasts who had learned to appreciate Adirondack parklands as campers, hikers, hunters, and second-home owners. Despite the state's longstanding commitment to promoting tourism and outdoor recreation, many Adirondackers still stigmatized the government as anti-economic growth for its often spotty enforcement of "forever wild." During the late 1960s and 1970s, the state began mulling over a new land-management regime for both state and private parklands, and, as a result, that stigma became a sort of self-fulfilling prophecy in the minds of many Adirondack residents who felt economically and politically disenfranchised.

CHANGING COURSE: THE TEMPORARY STUDY COMMISSION ON THE FUTURE OF THE ADIRONDACKS

In Albany, the time seemed ripe to strengthen the state's management authority over the Adirondack Park. By the late 1960s, the environmental movement was in full swing, and governments across the country began to play a more prominent role in managing lands beyond urban spaces and national forests. The practice of zoning lands by use originated in early twentieth-century cities, beginning with Los Angeles in 1908. Though first enacted at the behest of homeowners, real-estate firms, and merchants who wanted to segregate the landscape as a way to protect property values, urban planners embraced the practice as a method of rationalizing the cityscape. By the 1920s, zoning had reached the suburbs—but in the context of post–World War II sprawl, laws were not stringent enough for champions of open space.[26] Governments across the country responded. For instance, in order to curtail development pressures around the shores of Lake Tahoe, in 1960 California and Nevada created the Tahoe Regional Planning Commission (which would lead to the creation of the Tahoe Regional Planning Agency ten years later). The following year, Hawaii took an unprecedented step when the state adopted the nation's only statewide zoning plan in response to residential development of agricultural lands. Though no other state would enact a comprehensive land-management scheme like Hawaii's, over the course of the 1960s and early 1970s other states across the country would put into effect site-specific plans to protect coastal zones and other ecologically sensitive areas.[27]

Meanwhile, activists in the national "back-to-the-woods" movement pushed the federal government to enact what one proponent called "wilderness zoning." In 1964, they got their wish when President Lyndon B. Johnson signed the Wilderness Act, which established a procedure by which Congress could designate spaces "where earth and its community of life are untrammeled by man, where man remains a visitor."[28] Under the act, the National Park Service and U.S. Forest Service took on the task of mapping the lands under their purview to ensure that land-use patterns were consistent with development. Wilderness areas were supposed to be free from extensive and permanent modifications, and guests were supposed to leave no trace of their presence—the assumption being that for human beings to leave their mark was to spoil the environment. One Adirondack camper put it this way: "If you carry it in full, carry it out empty."[29]

In this context, New York State lawmakers and planners argued that the Forest Preserve in the Adirondacks needed more careful planning and stronger oversight. They decided that only human intervention could save the wilderness. Toward that end, in 1967, Laurance Rockefeller, Robert Moses's successor as chairman of the State Council of Parks and brother of Governor Nelson Rockefeller, commissioned a study that proposed that 1.7 million acres of the Adirondack Park's six million acres be converted into a national park.[30] The plan elicited near-universal scorn, particularly among Adirondack residents resistant to federal control and threats to their hunting rights; state administrators who wanted to maintain their authority; environmentalists who wanted to keep the state's stricter land-use controls; and the many New Yorkers who took pride in their state park.[31]

In the wake of Laurance Rockefeller's defeat, in September 1968 Governor Rockefeller tasked the Temporary Study Commission on the Future of the Adirondacks with devising a comprehensive plan that embraced the entire Adirondack Park as a unique state creation and natural environment. Rockefeller's group comprised mostly wealthy businessmen, like chairman Harold K. Hochschild, and political figures, like former New York City congressman Leo W. O'Brien, who knew the Adirondack Park as visitors rather than as full-time residents. As such, officials and business leaders in the Adirondacks feared that the commission would emphasize conservation over economic development. Hochschild, who owned a second home in the Adirondack Park, seemed to fit the old mold of the genteel Association for the Protection of the Adirondacks. And O'Brien's words confirmed many Adirondackers' worst fears: "Progress itself has produced a crisis—progress in the form of better highways which have brought the Adirondacks closer to large centers of population—and progress in the nature of more leisure time and financial resources for people who seek temporary surcease from urban environment."[32] In other words, the state now had to deal with the consequences of its successful transformation of the Adirondack Park into a modern wilderness playground for the masses. Toward that end, the commission completed detailed surveys of the park with the goal of formulating recommendations on how best to manage its expansive patchwork of public and private lands. In summer 1969, the commission began to hold hearings throughout the state to gauge public opinion on the park's future.

Many speakers at these hearings wanted to open up the Adirondacks to more recreational development for the benefit of vacationers and the region's businesses. Given the region's dire economic straits, restrictions on further development seemed like nails in residents' coffin. The 1960 unemployment rate in the Adirondacks was 11.2 percent, more than double the state's rate of 5.2 percent; and the region's economic outlook had not improved by decade's end.[33] Joe Uzdzvinis, a resident of Old Forge and former president of the company Central Adirondack Attractions, asserted that tourism was the means by which the greatest number of people could enjoy this state-protected natural treasure.[34] William Fountain, the supervisor of Essex County, expressed the "hope that the ... Study Commission ... will not be swayed by a relatively small but very vocal group of 'Forever Wild' enthusiasts.... We believe that the time has come to provide recreational facilities for all of the people of this state."[35] Fountain had to answer to his constituency, and since tourism was by the 1960s well established as a crucial industry in the chronically depressed Adirondack region, he appreciated the economic potential of developing both state and private lands. In the words of Claude J. Clark, an attorney representing the Franklin County Board of Supervisors, "we think in Franklin County that our future lies in the development of additional recreational facilities."[36] According to these proponents of mass recreation, democratic access to the Adirondacks best served the economic interests of locals and the recreational pursuits of tourists. In essence, they were championing the escalation of state conservationists' decades-long effort to make the Adirondack Park useful for the greatest number of people.

Development of the Forest Preserve, however, had changed the character of "forever wild" lands in a way that alienated the growing number of recreation seekers who sought out spaces that did not resemble the modern homes from which they were escaping. Members of increasingly vocal and influential environmental groups also spoke up at the hearings. State-driven recreational development, they argued, was having a material effect on what Margaret Lamy called the Park's "wild character." But what, precisely, did wild character mean? According to Frank J. Oliver, the president of the Adirondack Mountain Club, "the beauty, charm and stillness of the wilderness may be savored most fully with least damage to natural values."[37] Oliver's assertion turned the argument for mass recreation on its head, contending instead that all had the right to enjoy peaceful solitude in nature that overdeveloped campsites did not provide. Although these wilderness advocates likely did not represent a majority opinion in the Adirondacks, their forceful words stressed how much the places where they lived and played were becoming unrecognizable to them.

According to wilderness enthusiasts, the facilities provided by the state deprived the natural environment of its scenic beauty and appealing ruggedness. Since the 1920s, state administrators had, through their words and deeds, redefined "forever wild" to include the presence of extensive recreational infrastructure. During the 1960s, however, wilderness advocates sought to remove elaborate facilities from that definition. Alfred S. Forsyth of the Sierra Club argued, "conveniences . . . only debase and spoil the real attractions of the area which are its beauty, its grandeur, its quiet, its change-of-pace from city life; in short, its ability to recreate. . . . The word 'development' should be suspect. The area is already overdeveloped. . . . The real recreational values, and also the most powerful attractions for a paying tourist trade, lie in the beauty of the mountains, not in 'facilities' and the State should work to encourage the visiting public to recognize and appreciate these values."[38] According to environmentalists like Forsyth, proper camping entailed a return to simplicity and a deep connection with natural surroundings set apart from human-built structures. In their view, recreational facilities marred the landscape and diminished the outdoor experience.

As many observers noted, recreation seekers' marks on the land were becoming more conspicuous during the 1960s. By building up park infrastructure and employing rangers to clean up after campers, the state had endeavored to shield park-goers from the ambivalent consequences of their interactions with the land. However, the urban-style problems of overcrowding, unsightly development, and pollution were becoming increasingly difficult to ignore. The presence of large numbers of people in the Adirondack Forest Preserve, according to park resident M. I. Walsh of the Adirondack Wilderness Camp, was antithetical to wilderness: "small deposits of orange peels and gum wrappers are hidden under every convenient movable rock, and old burnt-out fireplaces are left beside brooks and ponds, or burnt into the duff. Walking through the woods it seems that one can never escape the refuse that campers have left scattered behind them in the wilderness."[39] Karl Gruppe, a member of the Izaak Walton League and Sierra Club, recalled more than one camping excursion during which he collected enough trash to fill up his canoe and car.[40] Harold Klein, a nature enthusiast from New York City, lamented the degradation of Follensby Pond, the very spot that, in 1940, seasonal resident Katherine Newbold had warned would become a danger to public health if it was overdeveloped. According to Klein, "The areas around these campsites were generally cleared of trees and had a well-worn look, the obvious result of much intensive use.... At least one of these camps stands out in my memory as a horrible eyesore. There was an assortment of junk scattered about the area: old iceboxes, washtubs, etc."[41]

Not only were these wilderness advocates questioning the value of mass recreation and challenging the utility of development, they were also subverting a core principle of the Conservation Department's mission. The department had conceived of building up the Adirondacks' recreational facilities as a way to cut the baby in half: to conserve natural resources while simultaneously making them useful. For environmentalists, however, the mass recreation encouraged by the state did little to conserve the environment—on the contrary, it seemed to promote destructive uses of the land.

Further, environmental activists argued that there were activities appropriate to the Forest Preserve, and they did not resemble those performed in modern cities and suburbs. Lysle W. Morton, the president of the Lake George Association, a state agency created to ensure the salubriousness and scenic value of Lake George, asked the Temporary Study Commission, "Will 'people who seek temporary surcease from urban environment' enjoy finding little Coney Islands and Atlantic Citys [sic] dotting the Adirondacks?"[42] New Yorkers living in more developed parts of the state, who knew the Adirondack Park as a recreational space rather than as a home, wanted to preserve places that did not resemble the cities and suburbs where they lived and worked. According to Per O. Moberg of the Long Island Environmental Council and Nassau County Fish and Game Association, "We, the metropolitan area residents, desperately need an area which is left undisturbed by man. We have readily within our reach all the man-made recreational amusements this civilization can dream of."[43] As a Long Island resident, Moberg was no stranger to the problem of vanishing open space. Large crowds, concrete-covered landscapes, litter, water and air pollution—urbanites and suburbanites traveling to parks to escape these undesirable features of their neighborhoods found them where they least expected.

As a result, the Adirondack wilderness seemed to be in danger of becoming too much like the urban landscape that many assumed rested on the opposite end of the environmental spectrum. Stanley Countryman of the Tramp and Trail Club in Utica testified, "We are against [more campsites], but we realize that they are inevitable but advise that the facilities now in use do not increase, and by 'facilities' I mean furnishing them with lights, and so on, and toilets, and so on, so that we will not have any further *suburban sprawls* within the campsites."[44] Countryman's choice of words was revealing. He articulated what he saw as a problem in the Adirondack Park—the loss of open space as a result of excessive recreational development—in terms of a larger environmental issue: the suburban sprawl that was enveloping so much of the countryside surrounding the state and nation's cities. The cities of tents that state planners had created during the interwar years became sprawling campgrounds, mirroring land-use practices being implemented outside the park. Clearly, the blue line surrounding the Adirondack Park on maps did not make the Forest Preserve impervious to the larger developments occurring outside its boundaries.

The Temporary Study Commission attempted to address wilderness advocates' grievances, and in its 1970 report recommended a new policy for campsite construction in the Adirondack Park. According to the commission, "The emphasis in new campsite location should be on dispersing use and on concentrating large modern campsites on the fringes of the Park; ... Only small rustic campsites should be built in the central Adirondacks; ... New campsites should be constructed only on land classified as wild forest land"; and "Campsite construction in the forest preserve should be in accord with the wild forest environment and with modern site planning principles."[45] The commission based these recommendations, which ran counter to five decades of state policy, on the belief that overconcentration, overdevelopment, and inadequate planning had characterized campsite construction since the 1920s, when state agencies began reshaping the wilderness for the benefit of automobile-borne city dwellers. The problems that arose from recreational development of the Forest Preserve were becoming untenable and, the commission concluded, a new path must be followed.

The Temporary Study Commission also weighed in on the state ski centers, and its report urged New Yorkers to exercise caution when considering new winter-sports facilities. If the cautionary examples of the two Whiteface ski centers were not discouraging enough, the commission threw cold water on the hopes of skiing boosters when it observed, "it is clear that the Adirondacks are not in a favorable position for intensive ski development. The main ranges run in a northeast to southwest direction, exposing many slopes to prevailing winds or prolonged sunlight. Abrupt easterly faces, rock outcroppings, excessive grades and narrow valley floors also present problems. Snow conditions often proved less reliable than in competing ski areas. Conditions, therefore, do not permit the development of a ski resort industry comparable to that in the Green Mountains of Vermont." Unlike the planners of Marble and Little Whiteface, the commission acknowledged the limits of administrators' ability to exploit an intractable natural environment in the interest of large-scale recreational development. As a result of unfavorable geography and weather conditions in the Adirondacks, the commission recommended more state incentives for private ski centers and expansion of existing state facilities, rather than the addition of new sites. Any new skiing area, the commission asserted, would require careful study of its environmental effects, scenic impact, and economic value.[46]

The commission's recommendations signaled to many Adirondackers that the close cooperation between the state and local business interests was coming to an end. Although businesspeople may have considered the state an often incompetent and unreliable ally, Albany had long been committed to promoting tourism in New York's North Country. The state's seeming reversal alarmed a great many area politicians and businesspeople. Even more distressing to them was the rhetoric of environmentalists, which at times seemed to erase the people living in the Adirondack Park. Year-round residents' moral ecology rested on their pride of place and their embrace of home rule, and they interpreted the state's move toward strengthening environmental regulations as another limit on their ability to control their communities.

Residents' determination to manage their own affairs took on greater urgency during the early 1970s, when the state began considering a new zoning scheme for the Adirondack Park's private lands. While the "forever wild" provision had to an extent limited human-caused changes on public lands, only natural obstacles, the conservationist impulses of landowners, and a few local zoning laws restricted development on private lands, which consistently accounted for about half of the park's acreage, even as its boundaries expanded. Was it necessary, then, for private lands to come under the state's purview, as well?

Among the most significant questions the commission sought to answer was: "What measures can be taken to assure that development on private land is appropriate and consistent with the long range well-being of the area?"[47] Planners, politicians, scientists, and laypeople recognized the interconnectedness of the park's public and private lands. Waterways flowed from one to the other; wildlife ranged across parklands without regard for property lines; and development of an individual patch of land siphoned resources away from and spread waste to neighboring parcels. The influx of people on the Northway threw this reality into sharp relief. The Adirondack Park had always been a place for multiple uses, but significant problems arose when second-home construction on an unprecedented scale highlighted the contradictions inherent in the development of a park shared by full-time inhabitants, seasonal residents, and vacationers.

Since large-scale residential development threatened to make dramatic changes to the Adirondack landscape, some took the position of embracing urban planning as a way to maintain the park's wild character. Peter E. Kress, a Rochester resident with a summer home in the Adirondacks, suggested during a December 1969 hearing "that zoning either become or remain stringent or well policed." As a seasonal park resident, Kress had a clear interest in maintaining the natural appeal of the place where he vacationed but did not work. Second-home owners were not the only zoning advocates. Chester L. Burton, the president of Paul Smith's College, spoke on behalf of the school's staff and faculty: "land-use zoning is necessary. Compatible uses should be grouped for the greatest good to the greatest number." Year-round Long Lake resident Orlando B. Potter wrote to the *Adirondack Daily Enterprise*: "Zoning offers the opportunity for protecting the public interest at minimum cost."[48]

Quite presciently, however, Harold A. Jerry, the commission's executive secretary, warned of a backlash against stringent land-use restrictions: "Many resident landowners in the Park will oppose the zoning of their land." For his part, John Stock, a Tupper Lake forester and commission adviser, asserted, "Development is of economic benefit to local residents. . . . There are a lot of people who feel that zoning is unamerican. The Adirondacks belong to the residents of the region." Adirondackers' resistance to state conservation policies since the late nineteenth century lent credence to Stock's admonition. Still, Jerry, along with commission chairman Harold K. Hochschild and the majority of members, believed that zoning was necessary. "There are only 103,000 residents of the Adirondack Park," Jerry observed. "Even if they all opposed zoning—and they don't—the interests of the remaining 18 million people in the state must be considered. Without effective planning and zoning, *their* Park will disappear."[49] Though the park would persist as a legally protected space, Jerry worried that a particular vision of the Adirondacks—one rooted in the recreational uses begun during the mid-nineteenth century—would vanish. Increased human pressures resulting from widespread second-home construction would significantly reshape the environment—thereby altering the nature and scale of the recreational uses in the park.

Among the Temporary Study Commission's 181 recommendations was the creation of an Adirondack Park Agency "with planning and land use control powers over private land in the Park." The commission suggested a nine-member, bipartisan agency, with seven private citizens, three of them legal residents of the park, to be appointed by the governor. They would be joined on the Adirondack Park Agency by the commissioner of environmental conservation and the director of the office of planning coordination. The APA would be tasked with devising plans for the entire park—subject to executive approval in the case of state lands and subject to both the legislature and governor's approval in the case of private lands. The agency's jurisdiction over private lands turned out to be the most controversial proposal. Stock's observation, quoted above, that many considered zoning "un-American" predicted the depth of the vitriol that would follow. The commission nonetheless sought to justify its recommendation: "while the state owned forest preserve of 2,250,000 acres is protected by Article XIV, unguided development on the 3,500,000 acres of private land will destroy the character of the entire Park if immediate action is not taken." The checkerboard pattern of land ownership in the park that was the product of decades of state action and inaction had intertwined the fates of public and private lands. Six hundred twenty-six owners held 53 percent of the park's private land, and each of those owners' individual tracts exceeded five hundred acres in size. Though at the time of the commission's report those landholders' property remained mostly forested, there were few restrictions limiting what they, and future owners, could do with their lands.[50]

Most distressing, according to the commission, was the possibility that those lands would become sites of massive residential developments. The commission observed, "The number of signs along the highways advertising new seasonal home and commercial developments attests to the growing pressures from an increasingly affluent and leisure oriented society." As a result, the commission warned, "without adequate planning and land use controls, subdividing will destroy the Park." The commission argued that small towns lacked the wherewithal to enact planning schemes for the good of their localities, region, and state, and thus required the resources, guidance, and enforcement powers of a state agency. Local lawmakers left to their own devices, the commission pointed out, had zoned less than 10 percent of private lands in the park.[51] The development of the Adirondacks, and the potential for further suburbanization, required, in the commission's view, an urban planning response with the APA.

THE BIRTH OF THE ADIRONDACK PARK AGENCY

The APA's birth was not guaranteed, especially once resistance emerged among lawmakers and Adirondack Park residents. Representatives of the region, particularly Republican state senator Ronald Stafford and Republican assemblyman Glenn Harris, worked to delay and, if possible, prevent passage of the APA legislation. Local opinion was divided, but generally opposed to the APA. The *Plattsburgh Press-Republican* argued: "the heart of the matter here really involves the ability of local people to govern themselves, to foster programs that they themselves feel are best for a single community." The supervisors of the towns of Newcomb and Glens Falls, and the legislatures of Essex, Clinton, Warren, and Washington counties, all recommended a year's delay in considering the APA legislation. The Adirondack Park Association, a conservation group that tended to favor recreational development, also advocated for postponement. Frank Casier of Saranac Lake, the owner of a furniture and appliance store and real estate developer who would become one of the agency's most outspoken critics, declared, "The state recommendations when read in their entirety are a plan to keep the average citizens out, to convert the Forever Wild Adirondacks into a Forever Empty Park, thereby defeating the purpose of having a Park."[52] The state's plans for limiting development fed the worst fears of local lawmakers and those residents concerned with the loss of control over their communities' resources and the worsening of their region's economic woes.

Their fears were not unfounded. APA opponents could recall the grassroots resistance to state conservation policies of the late nineteenth and early twentieth centuries. Whereas from the 1920s into the early 1960s the state and business interests worked together, if not always harmoniously, to construct elaborate recreational facilities, the struggles of the Whiteface ski centers and the defeat of the Hoffman Mountain amendment soured many Adirondack Park residents toward the state. Business owners, developers, and struggling Adirondackers were especially antagonistic, as the consistently bleak economic outlook for many was easy to connect with environmental protections (regardless of the validity of that connection). As Casier suggested, a park required a human presence to function as a park. Complicating matters, though, was the fact that the Adirondack Park, unlike places such as Central Park and Long Island's Jones Beach, had people living in it.

As a consequence of the Adirondack Park's unique mixture of public and private lands, the debate over the APA hinged on more vexing questions: *Who* should use the park? and *How* should the Park be used? "Average citizens," according to Casier, were the victims of the state's plan to limit construction on the Park's private lands. But who was the "average citizen"? Did Casier qualify? As a business owner and land developer he enjoyed more wealth than most in New York's North Country, and he had a clear economic self-interest in bringing more people into the region.[53] Yet he claimed to champion the right of ordinary people to enjoy the park—presumably as year-round inhabitants, seasonal residents, and recreation seekers. His use of the word "citizen" rather than, say, "resident" was revealing. Casier and other APA opponents claimed to be defenders of Adirondackers' rights as citizens, particularly liberty and the pursuit of property. However, their conception of the Adirondack Park as a space for mass commercial and recreational uses was at odds with APA supporters' desire to preserve the park's wild character.

Thus, opinion was not unanimous. The *Lake Placid News* argued: "We have not guarded or carefully watched our own impact on this area. . . . The clutter of private and public patches means that whatever private owners do affects far more than their own property. . . . We must recognize that our place is a very special, very unusual place in the world and we should do everything to keep it that way." William Roden, a newspaper columnist and former member of the now defunct Adirondack Mountain Authority, wrote, "All my life I have felt closer to the independent philosophies of the Adirondack natives than I do to the mass thinking of the cities.... But if it is necessary for such controls in order to preserve the Adirondack good life and I believe it is, then I will accept them." William and Janet Rochow, second-home owners with property on Tupper Lake, wrote the governor, "We welcome the proposals for more 'control' of the private lands." Plattsburgh resident Everett A. Maybell pushed Albany to act quickly "because real estate people and others are rapidly moving into the Adirondacks and developing private lands adjacent to those owned by the State in ways which are poorly planned and definitely injurious to the neighboring lands owned by the public." The Association for the Protection of the Adirondacks insisted that the APA Act "deserves wide bipartisan support." The *Essex County Republican* and *New York Times* urged swift action.[54]

One must also ask whose interests were APA supporters protecting? Second-home owners like the Rochows had a clear motive in saving the natural features that had brought them to the Adirondack Park as part-time residents. However, the vast majority of year-round Adirondack residents were not as privileged as vacation-home owners. Roden's argument that the APA would maintain the "good life" in the Adirondacks assumed that everyone benefited from the status quo. By contrasting Adirondackers' "independent philosophies" with urbanites' "mass thinking," he seemed to be harking back to the Jeffersonian ideal of self-reliant yeomen that had no basis in late twentieth-century reality. Such a perspective disregarded the impoverished Adirondackers who struggled to survive in the tourist economy, and for whom the promise of development had tremendous appeal. Yet not only were private lands at stake. Recreation seekers in and around the park, like Maybell, feared that "forever wild" lands would suffer as developers built up their properties. The state had made public lands accessible to motorists, enabling many average citizens to enjoy those wild spaces. By contrast, exclusive vacation-home communities did not benefit ordinary campers, hikers, and picnickers. APA opponents, it turned out, did not have a monopoly on populism.

As public debate raged on, Governor Rockefeller introduced the APA bill to the legislature on May 10, 1971, and it became law on June 7, with North Country lawmakers united in opposition. They were able to win concessions, however. The first was that the number of Adirondack Park residents on the agency was increased from three to four. A second and more significant amendment stipulated that, during the time between its creation and Albany's approval of a private land plan, the APA's interim powers prohibited it from exercising authority over towns that had passed their own zoning laws by July 1, 1971. The APA, which went into effect on September 1, was required to submit a state land plan in 1972 and a private land plan the following year.[55]

Since the new guidelines governing state lands required only executive approval, they went into effect with little fanfare. The APA submitted them to Governor Rockefeller in June 1972, and he promptly approved them. The Adirondack Park State Land Master Plan classified the Forest Preserve into seven categories distinguished by prescribed uses: wilderness; primitive areas; canoe areas; wild forest areas; intensive-use areas; wild, scenic, and recreational rivers; and travel corridors. The APA's definition of wilderness mirrored the language of the federal Wilderness Act: "an area where the earth and its community of life are untrammeled by man—where man himself is a visitor who does not remain." In these spaces, so-called "non-conforming uses" were prohibited, and even existing facilities, including tent platforms, boat docks, and truck trails, were to be dismantled. Only limited improvements and paths would be allowed to remain in primitive areas. On the other hand, wild forest areas like Hearthstone Point and Fish Creek Pond allowed for more intensive recreational uses.[56]

After the state had for decades domesticated campgrounds for the comfortable use of people temporarily leaving cities and suburbs, visitors who knew the Adirondack Park primarily as a recreational landscape demanded that its wild character be restored. In the process, they redefined "forever wild" once again, this time to mean the absence of *both* extractive industry *and* elaborate recreational facilities. Administrators responded by more thoroughly mapping and managing the land to control development, which involved separating wilderness from intensive-use areas in an attempt to sustain a multiple-use landscape that balanced competing human interests. High modernism could not get much higher. But there was still the matter of winning approval for the Private Land Use and Management Plan—which proved to be a far more daunting challenge for the APA and its supporters.

CONCLUSION

The creation of the APA marked a significant shift in state policy during the frenzy of the 1960s and 1970s. In the Northway dispute, "forever wild" purists had been willing to sacrifice private lands to save the Forest Preserve. The completed highway ended up having a significant impact beyond the state lands it traversed, however, as it encouraged more development throughout the Adirondack Park. To accommodate the people pouring into the region, the state continued, until the APA's master plan went into effect in 1972, its extensive development program on state forestlands. On private lands, meanwhile, real estate firms began to propose massive second-home communities to capitalize on the park's increased accessibility.

With the Adirondack Park under mounting pressure, New York State grasped a moment of possibility and implemented new policies governing land use. But attempting to enact a new scheme that ran counter to decades of established practice turned out to be a herculean task. Despite the support of many recreation seekers, second-home owners, and environmentalists who treasured parklands for their wild and scenic qualities, the APA would find itself constantly under attack.

As we will see in the next chapter, the most pressing challenge for the Adirondack Park Agency would be enforcing a comprehensive zoning ordinance for the park's private lands. The creation of the APA did little to stifle debate; in fact, conflict escalated along with the demands placed on the fledgling agency. More than a dozen different developers had filed plans for more than eighteen thousand houses in the Adirondack Park by the time the APA was ready to begin work.[57] And in response to those real estate firms' proposals, defenders of "forever wild" abandoned their willingness to sacrifice private lands at the altar of progress. The race for open space had begun: would the APA build a barrier on the track before developers reached the finish line?

CHAPTER 6

"THERE WAS ONCE AN ADIRONDACK PARK"

The Struggle over the Exurbanization of the Adirondack Park, 1971-1980

Sometime in the late twentieth century, Paul Bunyan took time away from lumberjacking to write a letter to Virginia, in which he lamented: "Yes, there *was* once an Adirondack Park." When and how did the park disappear? Bunyan had the answer: "1972 was the year a firm called Horizon . . . invaded the Adirondack Park. . . . In a few years, bulldozers swarmed *all over* the park and there was no forest left but that within the walled estates—which won't do *you* any good, Virginia."[1] A group called Citizens to Save the Adirondack Park, not "Paul Bunyan," authored this letter as a warning against unchecked development on private lands inside the Adirondack Park. The group's members joined together in 1972 to challenge Horizon's plan to build a massive second-home community in St. Lawrence County. They believed the fate of the whole park was at stake because real-estate developers had the capacity to make changes to the environment that dwarfed any made by campsite development and ski-center construction.

After decades of infrastructure improvement and recreational development, the Adirondack landscape reflected a dynamism—both ecological and human caused—that belied the notion that the park was forever wild. The struggle over land use in the Adirondack Park was fundamentally altered in the decades after World War II—at a time when standards of comfortable and healthful living in suburbs improved, second-home ownership expanded among the growing middle class, mass recreation reached new heights of popularity, and deindustrialization in both urban centers and the countryside left many Americans searching for new ways to make ends meet. Once the Northway increased access to the park during the late 1960s, the specter of large-scale residential and recreational development promised to bring about significant economic and environmental changes.

The planned vacation communities, in many ways, represented the next evolutionary stage following campsite improvement. The development of recreational infrastructure since World War I had been an attempt to widen the appeal of camping to the middle and working classes by making campgrounds more like home: comfortable, convenient, and healthful. While state lands underwent domestication, there were some who believed that the park's private lands had not reached their full productive potential, especially in the context of the chronic unemployment and poverty suffered by so many Adirondackers. During the late 1960s and early 1970s, real estate firms bought up land in the region—often from timber companies and other dormant industrial firms—in an attempt to capitalize on the expansion of home and car ownership among the middle and working classes that had created the potential for mass recreation at state campgrounds.

Leading the way were Horizon and Ton-Da-Lay, two companies that endeavored to provide actual homes—not just places *like* home—for the upper middle class. These vacation homes, like campsites, would be seasonal dwellings situated near scenic and recreational features. Developers conceived of second-home communities as recreational sites similar to Whiteface Mountain and its environs. In addition to houses, plans called for motels, shops, restaurants, trails, golf courses, ski centers, and lakes for swimming and boating. The accommodations in these second-home communities, of course, would be even more private, comfortable, modern, and durable than those found on campgrounds. These seasonal exurban enclaves would make up parks *within* the Adirondack Park.

Proposals involving the subdivision of tens of thousands of acres in the Adirondack Park sparked a bonfire of controversy. Opponents, who included mostly second-home owners, middle-class park residents, and environmentalists, worked to defend *their* wilderness: the rugged and scenic places where they played or lived part of the year. The large-scale residential development of private lands, activists argued, threatened the entire park—not just the lands on which vacation communities would be built—and they appealed to the fledgling Adirondack Park Agency (APA) to aid them in their effort to prevent the Adirondacks from becoming an exurb of surrounding suburbs and cities. Increasingly, both environmental activists and the state saw the Adirondack Park holistically, with the health of its public and private lands closely intertwined and interdependent. Accordingly, based on the conviction that the park's wildness and scenic beauty defined its character as a space for outdoor play, they sought to protect even private lands that were not legally designated "forever wild."

Environmentalists' opposition struck those in favor of economic development as a self-serving attempt to strip locals of the power to control their property and their ability to earn a living. Prodevelopment lawmakers and residents marched under a populist banner, accusing the state and its allies of implementing such stringent controls that the Adirondack Park would be kept as a playground exclusively for the wealthy. Although the well-heeled Association for the Protection of the Adirondacks would take a backseat to broader-based environmental organizations during the quarrel over the park's potential exurbanization, the image of the elite, genteel defender of "forever wild" continued to loom large in the debate. Prodevelopment Adirondackers joined forces with powerful corporate interests to champion the downtrodden against what they considered to be an oppressive and impersonal state bureaucracy. Antistate resisters even appropriated the language and history of past freedom movements, including the American Revolution, the abolition of slavery, and African American civil rights. They also drew on a regional tradition of resistance against conservation policies dating back to the late nineteenth century, as they formed a not-so-silent majority that opposed environmental regulations as state overreach.

One must be careful not to overgeneralize and oversimplify. Examining wilderness politics at the state and local levels reveals a story of changes, complexities, hypocrisies, and unexpected alliances not found in the literature on the national movement. Both sides in the debates over Horizon and Ton-Da-Lay claimed to have the best interests of both the natural environment and Adirondack Park residents at heart. Boiling the conflict down to environmentalists versus developers would obscure the ways activists in opposing factions marshaled competing ecological, economic, and class-based arguments. Even within the region there were a variety of responses to second-home construction that promised to increase the power of outside capital and bring tens of thousands of seasonal residents into small towns. In fact, the responses to Horizon and Ton-Da-Lay were significantly different. The Horizon Corporation, which planned a vacation enclave for up to 36,000 residents, did much to alienate Adirondackers and inspire spirited grassroots resistance. For its part, Ton-Da-Lay, Ltd., whose proposed second-home community would have increased the population of Franklin County from 44,000 to 64,000, gained much support in an area with a long history of antistate resistance. Park residents were not a likeminded mass. Class differences, economic self-interest, property ownership, and place of residence within the blue line that surrounded the Adirondack Park on state maps yielded a complex set of responses to developments that promised to transform Adirondack communities.

Though changes that took place during the 1960s and 1970s brought the clash over land use in the Adirondack Park to a head, history loomed large in the debate. The state's record of exercising extraordinary powers in the region; the decades-long shift to a tourist economy that created a dependent relationship between countryside and city; the long-held concern for the conservation of natural resources; and Adirondackers' proud traditions of self-reliance and antistate antagonism—all built toward the late twentieth century's battle of the North Country.

THE HORIZON CORPORATION AND THE RACE FOR OPEN SPACE IN THE ADIRONDACKS

As the Adirondack Park Agency was formulating its master land plans for the park, the first real estate firm to attract widespread attention (and inspire a mournful letter by "Paul Bunyan") was the Horizon Corporation of Tucson, Arizona. In spring 1972, Horizon purchased, from the Northern Lumber Company, 24,345 acres in the St. Lawrence County towns of Colton and Clare. With this transaction, Horizon took advantage of the region's flagging industry to become the thirteenth largest private landholder in the six-million-acre Adirondack Park. Horizon's plans for its property, if implemented, would have had a dramatic social and ecological impact on the towns targeted for development. The developer intended to build a community with seven thousand homes, an eighteen-hole golf course, a resort motel, and a shopping center, host to some thirty-six thousand seasonal residents. Colton, where most of the Horizon land was located, had a population of fewer than twelve hundred people.[2] The scale of the corporation's plans revealed in starkest terms the high stakes in the debate over the APA.

Horizon CEO Sidney Nelson was confident that the state would allow his plans to go forward. The property was the ideal spot, he claimed, because it was located on the northwest periphery of the park, where logging companies had deforested large swaths of land and, as a result, the natural features there were not as striking as the scenery found in the more popular Central Adirondacks and High Peaks areas. In the words of Horizon ally Frank Little, the owner of five St. Lawrence County newspapers, the development "would take over what is now wasteland, producing nothing, and make it into a multimillion dollar source of new income, new business, new employment and new taxes."[3] According to the pro-Horizon camp, now dormant extractive industry had left this corner of the park unattractive and unproductive, and it was up to real estate developers to invest the tract with both use and monetary value.

If Horizon and its supporters expected nature enthusiasts to ignore the project because of its location, however, their hopes were quickly dashed. Opposition to Horizon coalesced around the grassroots organization Citizens to Save the Adirondack Park. Evelyn Cropper, a founding member and a resident of Canton, a college town about fifteen miles outside the park, spread the word that the corporation "seek[s] to get bulldozers into the Park before the newly created Adirondack Park Agency has time to create and reveal its master 'land use and development' plan." Horizon was just the opening wedge, she warned. Cropper's group was joined by, among others, the New York State Conservation Council, St. Lawrence County Environmental Management Council, Students for Environmental Alternatives, and the many unaffiliated Adirondackers who flooded their local newspapers and state offices with letters. By July 1972, Citizens to Save the Adirondack Park, which at its height boasted twenty-six hundred members, had collected more than five thousand signatures on a petition against the subdivision.[4] The organization effectively tapped into strong antidevelopment sentiment both inside and outside the park (see figure 10).

Nevertheless, the rhetoric used by Horizon supporters relied on the often dubious notion that the company's detractors were outsiders stirring up trouble in the Adirondacks. Rather than debate the merits of the corporation's plans, it was easier to undercut the credibility of opponents by stigmatizing them as interlopers. Harold Buck, the Colton town supervisor who also directed a local funeral home, argued, "We can't stand still, can't stagnate," and derided Horizon's opponents as meddling intruders. When reminded that Scott McRobbie, an outspoken critic of Horizon, lived in Colton, Buck dismissed him as "one of the smart professors" because he taught physics at Potsdam State College. At the same time, though, Buck pointed out that Horizon "will attract a good class of people. The riffraff couldn't afford it." While impugning McRobbie as an intellectual elitist, Buck also welcomed an influx of wealthy part-time residents into his town. He was not alone. According to the owner of an area hardware store, "outside instigators

FIGURE 10. "Save the Adirondack Park," a poster by Robert Plumb for Citizens to Save the Adirondack Park, c. 1972. Citizens to Save the Adirondack Park commissioned this poster in their campaign against the Horizon development in St. Lawrence County. Courtesy of the Adirondack Experience and the family of Robert Plumb.

have upset local people. . . . They're making a good living in the city themselves, but they don't want us to."[5]

As we have seen, this love-hate relationship between Adirondackers and city folk had a long history. While local businesspeople and workers relied heavily on urban vacationers' money, they often resented this dependent relationship, especially when nonresidents attempted to direct policy in the park. The moral ecology embraced by many locals dictated that they should be the ones to determine land and resource use in their communities. However, the tourism industry promoted by the state, local governments, sports clubs, wealthy landowners, and private companies had given outsiders a stake in the Adirondack landscape—and, much to the chagrin of many Adirondackers, those seasonal residents and recreation seekers had been defending their own interests with the considerable resources at their disposal for decades prior to the clash over residential development. Ironically, at the same time that some Colton residents disparaged Horizon opponents as outside agitators, the project's supporters encouraged an Arizona-based corporation to clear thousands of lots for people who lived most of the year outside the park. Evidently, not all "outsiders" were created equal.

Despite the perception of Horizon supporters, much of the opposition came from people who lived in or near the park and wondered how Colton, a town of twelve hundred people, would cope with the thirty-fold increase of its population every summer. In fact, St. Lawrence County's own planning director, Richard Grover, predicted: "If the people come [to Horizon]—Colton is gone." Infrastructure like roads and sewage systems would require significant expansion, and services like garbage disposal and fire control would be taxed to an unprecedented degree. Scott McRobbie, for his part, went further and warned that property taxes would skyrocket out of reach of the "little man." In response to the aspersions cast by Horizon supporters, McRobbie asserted that he was not part of "a raving radical movement. It's just the people wanting to save a little bit of nature the way it is."[6] Although the Horizon property may not have been located in the most scenic area of the Adirondacks, the company's foes feared that such an immense project would both reduce locals' standard of living and irrevocably degrade the environment. As longtime Colton resident Lionel P. Hepburn wrote to the APA, "We have no need for services of far-off outsiders who invade our Township or County and the Adirondack Park with big ideas of over-riding the 'Forever Wild Laws' that were so prudently established many years ago."[7] Hepburn, like many advocates for wilderness protection, conflated state-owned woodlands with private parklands. Even though Horizon, in fact, planned to subdivide private lands not protected as "forever wild," Hepburn still considered the corporation to be an unscrupulous invader seeking to transform his home and destroy the recreational landscape. Where those in favor of the Horizon plan saw a wasteland in need of development, opponents saw an attractive and unified ecological system under threat.

Among locals, economic self-interest, class dynamics, and environmental consciousness, even more than place of residence, served to separate the opposing sides. Small business owners and many working-class and poor Adirondackers harbored bitterness toward Horizon opponents, whom they considered to be wealthy elitists telling them how to live. The anti-Horizon camp, for its part, used similar anti-outsider rhetoric to assail the company and claimed to carry the mantle of local economic and environmental interests against a developer encroaching on their homes. The imperative of home rule, which had been integral to Adirondackers' sense of moral ecology even before the creation of the Adirondack Park, guided Horizon's critics just as that principle buttressed the position of the company's advocates. Although both proponents and opponents of Horizon shared the same space, they embraced a different vision of place: the former envisioned a growing community with thriving commerce and the latter a small human presence surrounded by a scenic natural environment open to relatively low-impact recreational uses.

While the APA got its house in order, Horizon came under fire on other fronts. In October 1972, New York State fined a subcontractor employed by the corporation for trespassing and cutting trees in the Forest Preserve. When called before the state assembly's Joint Legislative Committee on Environmental Management in December, the company declined to send a representative. Even the assemblyman Glenn Harris, one of the APA's most committed adversaries and a promoter of development in the Adirondacks, blasted Horizon as a "corporation with a public-be-damned attitude" for its failure to appear. In 1975, the Federal Trade Commission filed a lawsuit against Horizon for alleged misleading advertising and building practices, and the following year Arizona and New Mexico sued the company.[8] As we will see, Horizon would be no stranger to New York State's courtrooms either. Before the company would see its day in court, Horizon sought and failed to win from the APA conceptual approval for its plans but never made a final proposal. While Horizon temporarily faded into the background, attention turned to another development in neighboring Franklin County, called Ton-Da-Lay.

CORPORATE LAND MANAGEMENT VS. STATE REGULATION: THE CASE OF TON-DA-LAY

In June 1970, real-estate developer Louis Paparazzo purchased for $1.5 million 18,386 acres in the Franklin County towns of Altamont (since 2004 Tupper Lake) and Santa Clara. The following year, in September, Paparazzo began courting local business interests with a speech before the Tupper Lake Chamber of Commerce. He touted his accomplishments during an almost thirty-year career, in which he and his family had built housing developments in Indiana, California, and Connecticut. He had saved his most ambitious plans for the Adirondack Park. Before a room of local business leaders, he laid out a plan to house up to twenty thousand people—more than three times Altamont's population of sixty-three hundred—in a second-home community called Ton-Da-Lay (a name he made up). This development would include four thousand homes and three hundred acres of resort and commercial space, centered on eight ponds and lakes. It would be a modern, multiuse recreational landscape featuring a camper area, an eighteen-hole golf course, ski centers, beaches, marinas, hiking, biking, snowmobile trails, bridle paths, and an equestrian space. Paparazzo promised affordable lot prices so that the middle class would be able to enjoy the amenities offered at Ton-Da-Lay.[9]

Houses and recreational facilities were not Paparazzo's only selling point, however, for he also promised to leave more than 90 percent of the land untouched. In fact, the tract's naturalness was part of its appeal. As Paparazzo put it, "The terrain is such with its mountains, hills, valleys, lakes and streams that it provides the scenic background for the fulfillment and enjoyment of vacation and leisure hours in a natural environment." Although the APA had been created less than three weeks before Paparazzo's speech, he felt confident that his project would proceed smoothly since Altamont had adopted its own zoning code in 1963, thereby temporarily exempting the town from agency oversight.[10] The APA was not even necessary here, the company asserted, because "Ton-Da-Lay, Ltd. is a responsible economic enterprise" whose efforts would have a positive impact on the natural environment.[11] At a time when environmental issues were the subjects of intense debate in the Adirondacks and elsewhere—the Ton-Da-Lay tract was purchased two months after the first Earth Day—Paparazzo portrayed himself as an ecologically sensitive developer.

Paparazzo pledged that an elected board of governors would prevent overdevelopment at Ton-Da-Lay by enforcing protective covenants. The board would be responsible for approving siting, land-clearing, materials selection, and construction, and ensuring that owners did not build on more than 20 percent of their individual lots. Some rules were intended to set Ton-Da-Lay apart as a superficially respectable, middle-class community: items like clotheslines and trashcans had to be shielded from neighbors' view, and lots could not have junk or idle automobiles. Seventy-five percent of Ton-Da-Lay residents would have to vote to approve any changes to guidelines. Essentially, Ton-Da-Lay's board of governors would function as a private, small-scale Adirondack Park Agency, and the protective covenants as a master plan restricting land use.[12] Regardless of whether the APA or Ton-Da-Lay had authority over those lands, owners would not enjoy unfettered independence to use their property as they saw fit. The choice with Ton-Da-Lay was not between freedom and state regulation; nor was it between individual rights and state power. Rather, it was between state control and corporate control.

Though exempt from APA review for the time being, Paparazzo would have to pass muster with the Department of Environmental Conservation (DEC). Established in 1970, the DEC absorbed the land- and resource-management responsibilities of the now defunct Conservation Department and many of the pollution-control duties of the state health department. According to the state conservation law, since Paparazzo proposed building more than fifty housing units on his property, Ton-Da-Lay would have to clear its water-system plans in a series of public hearings. Though Paparazzo had already jumped over several bureaucratic hurdles, receiving approval from the state health department, Franklin County legislature, and Altamont town board, he would have to prove to the DEC the healthfulness of his initial plan to subdivide 301 lots on one thousand acres.

Whether local authorities had performed their due diligence when evaluating Ton-Da-Lay's application was the subject of debate. When a reporter for the *Adirondack Daily Enterprise* contacted Carl Felton, the chairman of the Altamont planning board, the planner was uncertain about the specifics of the Ton-Da-Lay proposal and why it had received a thumbs-up. The chairman of Altamont's planning board, the reporter noted, did not even recall what lot size the town's zoning ordinance required. Whether Fenton's response was the result of ignorance or poor memory, his words did little to dispel the fears of state regulators and environmentalists that local planning boards either lacked the wherewithal to implement their own comprehensive guidelines or, worse, functioned as a rubberstamp for any developer that promised an economic boost in their towns. In fact, in 1963 the Altamont town board had adopted, seemingly inadvertently, an unfinished draft of its subdivision regulations and subsequently disregarded the consulting firm's final submission.[13] In part as a result of these missteps by local planners and lawmakers, Ton-Da-Lay's opponents saw the protective covenants as a fig leaf covering up Paparazzo's intention of squeezing every dollar out of the land by developing it beyond capacity.

What was at stake in the DEC hearings? According to some residents, their very survival hung in the balance. Franklin County was among the poorest Adirondack counties. Major area employers closed their doors and laid off hundreds of workers during the late 1960s and early 1970s. For instance, in 1967 Armour closed its meatpacking plant in Tupper Lake village (in Altamont) and moved its operations to Plattsburgh. Two years later, U.S. Plywood-Champion Papers shuttered its Tupper Lake factory, laying off one hundred and fifty workers.[14] Consequently, in 1970, Franklin County had New York's highest unemployment rate of almost 17 percent, and its residents earned the state's second lowest average annual income.[15]

Although Ton-Da-Lay would not house many native Adirondackers, locals hoped the project would employ construction workers, craftspeople, and service staff, as well as bring more customers for local businesses. The proprietor of the drugstore on Altamont's Main Street said, "We've got a bunch of super birdwatchers who are against Ton-Da-Lay.... We have to survive, and if they put a fence around us and say no development, we're in trouble." The genuine desire of locals for economic growth provided momentum for Ton-Da-Lay's plans. Others, however, feared the unintended consequences of development. One Altamont native, William L. Bentley, expressed concern that the town's residents were so preoccupied with "just getting along that they haven't got time to think of the consequences." The land would undergo substantial change if Paparazzo's plans went forward, but the repercussions were hard to predict. When one of Ton-Da-Lay's salespeople took a prospective buyer and a *New York Times* reporter on a tour of the tract, the journalist described it as a place "where the only signs of civilization were a few isolated hunting shacks and the loudest sound was the cry of a loon from a sparkling lake."[16] What would happen if "civilization" came?

The debate over Ton-Da-Lay was about more than a single development and town, for the realization of Paparazzo's plans would have had a significant impact on the neighboring Forest Preserve and potentially set a precedent for future development in the Adirondacks. When, in May 1974, Phil Gallos of the *Lake Placid News* reported on his visit to the Ton-Da-Lay tract, his description rooted the ongoing land-use debate in a tangible environment. At Dry Channel Pond, whose shoreline was slated for subdivision into dozens of lots, Gallos observed, "the shore front drops very steeply into the water and the ground appears to be of either esker or moraine structure (very unstable). To develop the area around the pond as planned is going to require a great deal of care and engineering expertise—and a healthy dose of common sense." After walking a little over a mile, Gallos entered an open meadow on public land and saw the very dry channel from which the pond got its name.[17] As he crossed back and forth between the Forest Preserve and the Ton-Da-Lay tract, Gallos highlighted how fragile particular natural features were and how closely intertwined private and public lands were, thus raising the question: was the Ton-Da-Lay plan compatible with both the environment and established uses?

In fact, during his trek through the Ton-Da-Lay tract, Gallos found that "civilization" had already arrived when he stumbled upon fishing boats and campgrounds with tents, fireplaces, and a garbage dump. "We shudder to think of what will happen to this place when the future residents of Ton-Da-Lay find out about" the trash heap, Gallos noted.[18] Even before construction began, the seeds of discord had been planted in the landscape. Paparazzo's parcel and the surrounding lands were not untouched wilderness. Rather, they were host to recreational uses that included hunting, fishing, canoeing, and camping. The debate over residential development, then, was not over *whether* the land should be used, but instead over *how* the land should be used. The Ton-Da-Lay tract bordered relatively undeveloped lands belonging to the conservationists of the Kildare Club, the Boy Scouts, and William A. Rockefeller, as well as the state-controlled Saranac Lakes Wild Forest and St. Regis Canoe Area. In fact, the terrain being contested during the 1970s was some of the same land at the center of the conflict during the first two decades of the Adirondack Park's existence.

It was in this context that the DEC proceedings, held in Ray Brook, began on December 5, 1972. In a controversial decision vigorously protested by Ton-Da-Lay's representatives, the DEC commissioner Henry L. Diamond declared, "It is our duty and responsibility to use the hearing process to develop a record of the full ecological impact of the proposed development rather than sanctioning ourselves to only the narrow and immediate limits of the water supply application."[19] The ensuing debate hinged on questions of state authority and the potential environmental impact of development.

Paparazzo, still confident despite the broad scope of the hearings, continued to speak environmentalists' language in his testimony. "Thinking has been to develop a project that will have men and nature and harmony," he said. Although he admitted that he could not guarantee his protective covenants would be followed, he assured skeptics that guidelines would be enforced for as long as he was alive. In his promise to bring ecologically sensitive development to the Adirondacks, Paparazzo appealed to residents' moral ecology. In the words of Ton-Da-Lay witness David E. Wood, president of the Franklin County

Chamber of Commerce, "We love the beauty of the Adirondacks as much as the outsiders who don't want this project but we are in favor of good clean development."[20] Ton-Da-Lay's supporters asserted that they appreciated nature as much as the project's opponents did, and they promoted what they considered to be a middle-ground approach intended to benefit both people and nonhuman nature.

Despite Paparazzo's often stated concern for his land's ecological health, the development he proposed distressed many nature enthusiasts. William Bentley, who, as a local Ton-Da-Lay opponent was far outnumbered by supporters at the standing-room-only DEC hearings, warned of what he called "the megalopolis suburban pressure syndrome." Libby Beyer, a Morrisonville resident who attended one of the early hearings, signed her letter to the *Plattsburgh Press-Republican* as "An Environmentalist," even though that label was a pejorative among many Adirondackers. According to Beyer, "this term means someone who cares enough about the world and mankind to seek to alert the many forces that would endanger both. The time will come when environmentalists will no longer be referred to as 'crazy bird watchers.'"[21]

According to Beyer and other Ton-Da-Lay opponents, the choice between what was good for people and what was good for the environment was a false one. They agreed that the Adirondack Park's uniqueness was rooted in its open space, beautiful scenery, and wildlife. Now the park was in danger of falling prey to sprawl, which, they argued, would have catastrophic consequences for both locals and nonhuman nature. The links between city and the Adirondacks were abundantly clear by the early 1970s. First rail lines, then roads, brought urbanites to the region, and the money they spent in the region was crucial to the economy of New York's North Country. Preservationists had warned of the overdevelopment of Forest Preserve campgrounds and Whiteface Mountain, and now environmentalists feared the exurbanization of the park's private lands. The threats were not theoretical or the products of a distant future, as places like Long Island served as cautionary examples.[22] It was this kind of pressure the APA was created to alleviate, but with Ton-Da-Lay the agency was powerless.

With land and money at stake, the rhetoric was heated from the start. Most alarmingly, David Vanderwalker of the Franklin County Board of Supervisors likened antistate resistance in the Adirondacks to the cause of eighteenth-century American revolutionaries, and warned that if the DEC deprived Adirondackers of the Ton-Da-Lay development, "We may see one of the biggest bonfires in history." Such language permeated much of the rhetoric for the prodevelopment side, as Ton-Da-Lay supporters went beyond celebrating the potential economic benefits of development and couched their position in American principles and rights. Indeed, Vanderwalker was harking back to the tumult over late nineteenth- and early twentieth-century conservation policies, which actually resulted in arson and other crimes.[23] As we will see, some Adirondackers would go to similar lengths to protest the state's expanding role in the park.

Not long after the DEC hearings on Ton-Da-Lay began, the APA added more urgency to the proceedings when, on December 21, 1972, it distributed the private land plan for comment. The agency proposed dividing private land within the blue line into six categories: (1) hamlet, where development would be concentrated; (2) moderate intensity use; (3) low intensity use; (4) rural use; (5) resource management; and (6) industrial. Each category would be governed by a set of guidelines restricting the density and placement of construction projects. The agency would be charged with evaluating "regional projects" (those that raised significant park-wide concerns); "special uses" (those that involved less substantial concerns than regional projects); and "permissible uses" (those involving local issues). The most restrictive category was resource management, which included "areas [in] which the need to protect, manage, and enhance forest, agricultural, recreational and open space resources is of paramount importance."[24] Most of the land in the Ton-Da-Lay tract fell under resource management guidelines, an ominous sign for Paparazzo as he raced to obtain approval from the DEC before the APA gained jurisdiction over his property.

While the DEC continued to hear testimony in the Ton-Da-Lay case, in January 1973 the APA held fifteen public hearings on its proposals. As the private land plan circulated, resistance among Adirondack lawmakers and residents grew. Richard B. Purdue of Indian Lake wrote to the *New York Times*, "Can you seriously justify the effort of the Park Agency to make this local government a slave to Albany?" Ticonderoga Fish and Game Club president Merrill Dolbeck argued, "As we continue to lose our authority, our God-given American and our Constitutional rights . . . we become the puppets and pawns for those whose selfish desire and aim is to control and levy." During one APA hearing, a man dressed in American Indian garb carried a sign that read, "Now you've taken the land, point the way to the reservation." At another hearing, Franklin County legislator John M. Sparks said to enthusiastic applause, "These [environmental] groups . . . do not give a hoot whether or not . . . the guy who owns one or two lots in the Adirondacks and who lives here year round and has to eke out his living by working day and sometimes night, lives or dies or has his land and his savings slowly taken from him."[25]

APA opponents took on the role of the oppressed minority as a way to assert their rights and resist state control. By referring to slavery and the oppression of American Indians, anti-APA activists placed the agency's measures in the context of state-perpetrated transgressions against Americans' freedom and well-being. Perhaps, too, they were asserting their own close ties to the land, which, they argued, outsiders could not possibly understand. To those who resisted the agency, environmentalists were enemies of the progress on which residents' survival depended. From APA opponents' point of view, local power and property rights were at stake, and to sacrifice them would be to surrender to an assault on their constitutional rights.

In response to criticisms expressed at the hearings, the APA submitted a revised plan to the legislature in early March 1973, but its path to passage was akin to climbing Whiteface Mountain on foot during a blizzard. Even though the revised proposal loosened restrictions on the density of construction—for instance, allowing fifteen principal buildings per square mile on resource management lands instead of ten—APA commissioner William J. Foley, a lawyer from Old Forge, issued a dissenting report in which he argued, "the plan constitutes an unconstitutional taking of the private owners' property." Adirondack town supervisors, for their part, mobilized to delay the bill, winning a temporary injunction to halt publication on the grounds that the APA had not solicited sufficient local input. In late March, both the state assembly and senate passed Glenn Harris's bill to delay consideration of the private land plan for a year. While the bill sat on Governor Rockefeller's desk, negotiations between the governor's office and the legislature continued behind the scenes.[26]

On May 9, Rockefeller vetoed the Harris bill and the two sides agreed on a compromise plan. Critics secured some two hundred changes to the Agency's original proposal. Development restrictions remained largely the same, though, with some revisions to the APA's guidelines for construction near highways and lakeshores. Extending an olive branch to the opposition, Rockefeller agreed to the addition of the state commerce commissioner to the agency. Also as part of the compromise, a twelve-member Adirondack Park Local Government Review Board appointed by county legislatures was empowered to observe and advise the agency, and report directly to the legislature as the representative of park residents. Though created as an advisory panel, the review board went on to become the APA's fiercest nemesis. With these and many other changes, the private land plan won majorities in both the assembly and the senate—without the votes of North Country lawmakers. After the bill's passage in the legislature, Glenn Harris said, "It doesn't bring an economic ray of hope to the area. It places a knife in the heart of it." On May 23, 1973, Rockefeller signed the Adirondack Park Private Land Use and Management Plan into law. It was set to go into effect on August 1.[27]

As the DEC hearings on Ton-Da-Lay dragged on and each step toward the approval of the APA's private land scheme ticked off the time Paparazzo had to implement the plans for his property, the developer's frustration boiled over. In press interviews, he acknowledged that his protective covenants would limit construction on the plots he sold, but noted that purchasers would know of the restrictive guidelines in advance. The state-enforced rules, on the other hand, would be imposed after he had purchased his property. He then ratcheted up his rancor: "never since my reading of Lennin [sic] or Mein Kampf have I read a better planned confiscatory document than the Agency's Private

Land Use Plan." He referred to Vanderwalker's controversial statement about bonfires and added: "Ironic that on the eve of our bicentennial celebration we should find ourselves fighting to protect those same rights.... [W]e have no bay for a tea party but we can light bonfires."[28]

Evident in such heightened rhetoric was the voice of the conservative resurgence in the 1970s and 1980s. (This language was later echoed in the rise of the Tea Party in 2009–2010; in fact, one group of APA opponents in the St. Lawrence County town of Fine called themselves the Minutemen.)[29] Central to that political shift was a backlash against environmental regulations that were interpreted as methods of stifling free enterprise and restricting use of private property. Adirondackers' and developers' assertions of their right to control the resources in their communities predated the Sagebrush Rebellion, during which, beginning with Nevada in 1979, state legislatures in the West voted to take lands from the federal Bureau of Land Management and promote more intensive development.[30] Hedging his bets, Paparazzo smeared the APA as both communistic and fascistic. State environmental regulations, in his view, were mechanisms for the subjugation and impoverishment of people. While he portrayed himself as the ecologically sensitive developer who would bring much-needed jobs to the region, he portrayed the state as *the other:* an intimidating, alien force attempting to impose un-American values on the Adirondack Park.[31] Thus, the state beat out environmentalists, city dwellers, and professors for the part of most threatening outsider in the APA resisters' casting call. Such language carried great weight among the many Adirondackers who prided themselves on their individualism and carried on a tradition of anti-government resistance dating back to the late nineteenth century.

Unfortunately for Paparazzo, he butted heads with a state environmental regime growing in strength thanks to widespread popular support, legislative initiative, and judicial sanction.[32] Enforcers of environmental policy in New York had a growing set of resources on which to draw when debating their prodevelopment foes. For instance, the Sierra Club, an environmental group that dated back to 1892 but had been enjoying tremendous growth in size and influence since the 1950s, intervened in the DEC hearings as a Ton-Da-Lay opponent. Not until the controversies over second-home development did national environmental organizations like the Sierra Club begin to play a leading role in the contests over land use in the North Country, signaling both the expanding reach of these groups and the significance of the Adirondack wilderness. Among the several experts the Sierra Club called to the stand at the DEC hearings was Philip Hoff, who as governor of Vermont from 1963 to 1969 had witnessed construction of some twenty-three thousand second homes in his state, which is roughly the size of the Adirondack Park. According to Hoff, after vacation-home construction provided an initial economic boost to the Green Mountain State, Vermonters suffered the consequences of what historian Hal Rothman called a "devil's bargain," and the crippling costs overrode the benefits. The price of land ballooned, and local governments struggled to maintain the infrastructure necessary to accommodate a significant influx of people. As for the promise of jobs, Hoff noted that second-home developments tended to create seasonal, low-paying service jobs. Increased property values in and around these second-home communities, he further testified, pushed poorly paid laborers farther away from their workplaces, negating to a degree their meager gains in income. He warned: "Wilderness areas are few and far between, and the pressures on them are enormous."[33]

Paparazzo's own history included worrisome signs that the precedent Hoff warned of might become reality in Franklin County if Ton-Da-Lay came to fruition. The Connecticut *Commercial Record* published a report on the impact of the Paparazzo family's Heritage Village, a community for senior citizens, opened in Southbury, Connecticut, in 1965. When Southbury residents voted to change the town's zoning laws to allow construction of Heritage Village, they hoped the Paparazzo-Heritage Corporation would foster economic growth and simultaneously take some of the tax burden off residents. New roads within Heritage Village were supposed to be the company's responsibility, and since children would not live there, tax money would not have to be spent on new schools. Sweetening the deal, Paparazzo promised to build an access road to the property if the town agreed to replace a crumbling bridge across the nearby Pomperaug River.[34]

However, according to Southburian "Mrs. George Nichols," "before the vote the town officials said it would be a wonderful thing and lower taxes. . . . In fact, with a new road to the village, it looks as if taxes are going higher than ever." As land values in Southbury skyrocketed, Paparazzo's company filed a series of lawsuits against the town, challenging assessments while withholding almost $100,000 in taxes. Meanwhile, as Heritage Village grew in size and became home to more voting inhabitants, its occupants used their clout to pass a resolution that shifted responsibility for the road away from the corporation and to the town. Short on revenue, Southbury was forced to spend $500,000 on both the bridge and access road. First Selectman Richard A. Harroff lamented the company's ability to "impose demands on this town which are often distorted" by "load[ing] a town meeting" and "raid[ing]" a political caucus.[35] Newcomers to Southbury gained a measure of influence the town's leaders and residents had not anticipated, and Southburians outside Heritage Village bore a greater financial burden as a result. The power of seasonal residents to shape policy in the Adirondack Park was compelling evidence that the newcomers who moved into Ton-Da-Lay would enjoy outsize influence to the detriment of year-round Franklin County residents. Though Paparazzo trumpeted his dedication to improving Adirondackers' lives, his opponents' skepticism was rooted in more than knee-jerk antagonism, for his family's record called into question the promised economic benefits of Ton-Da-Lay.

As Hoff's testimony and the story of Heritage Village suggested, besides the more abstract principles being debated in the Ton-Da-Lay controversy, there were pragmatic questions about what was really beneficial for the region's natural environment and people. The notion that the choice was between the environment and the economy was a canard that obscured more than it illuminated. As in the earlier debate over the Northway, both sides in the Ton-Da-Lay imbroglio claimed to be fighting for the best ecological and economic results. Paparazzo appealed to Adirondackers' moral ecology when he promised jobs, affordable vacation homes, *and* an environment made more healthy and healthful. On the other hand, Ton-Da-Lay's opponents pledged to save both the natural environment *and* Adirondackers from projects that they predicted would impose overwhelming burdens on the region's land and people. The conflict's outcome would be determined, of course, by power—that is, by which side was best able to marshal the financial, political, and legal resources to win the day. But it was also contingent on the ability of the debate's participants to ground their arguments in concrete reality. Though the prodevelopment faction tended to appeal to the ideals of liberty, the sanctity of private property, and the free market, and the opposing camp often rooted its cause in nature appreciation, the value of outdoor exercise, and the spiritual interconnectedness of living beings, the conflict over Ton-Da-Lay was also based on observable, measurable consequences.

The April 5, 1973, testimony of Richard F. Estes, assistant director for regional planning with the APA, underscored the different conceptions of appropriate land use in the Adirondacks embraced by the opposing sides. The APA, Estes explained, considered several different factors when assessing Adirondack parklands' capacity to sustain development: the quality of the soil, the land's slope and elevation, whether the area was under active timber management, the existence of critical wildlife habitat, the fragility of plant communities, and the presence of unique scenic and historic features. The Ton-Da-Lay tract contained many features deemed ecologically significant, including Iron Mountain, Spring Pond Bog, and a stretch of the Jordan River. Proximity to state lands was another factor—an especially significant one considering Ton-Da-Lay's closeness to the Saranac Lakes Wild Forest and St. Regis Canoe Area. As a result of Ton-Da-Lay's resource management classification, once the Private Land Use and Management Plan went into effect Paparazzo was limited to constructing fifteen principal buildings per square mile, or no more than thirty structures on the first one thousand acres. He planned to build 301.[36] Though at the time of Estes's testimony the legislature was still debating the private land plan, it would go into effect on August 1, 1973, the same day the conservation commissioner Henry L. Diamond rendered his decision to deny Ton-Da-Lay's water-supply application.

Diamond followed the recommendation of the hearing officer,

attorney Frank V. Carine, who, after considering 220 pieces of evidence and the testimony of thirty-five witnesses, determined that Ton-Da-Lay had failed to demonstrate that its plans would avoid adverse environmental impacts. Carine wrote, "Development of the magnitude envisioned will alter the species, composition and quantity of wild life in the area." As for the land and waterscape: "Effluent will be produced in large quantities and if it enters water bodies untreated, it will render them useless for bathing and, eventually, fishing." The waste resulting from the concentration of up to twenty-thousand seasonal residents in Ton-Da-Lay, Carine feared, would threaten recreation seekers' continued enjoyment of less intensive uses of public lands. As we have seen, the poorly constructed sanitary facilities at campsites clearly demonstrated the need for care when building sanitary infrastructure. In addition, Carine was not convinced by Paparazzo's assurances that the lands set aside as "wilderness" would be left undeveloped, because the protective covenants did not apply to those parcels. Finally, Carine pointed out, Ton-Da-Lay did not provide an adequate environmental impact assessment.[37]

Indeed, the document Ton-Da-Lay had submitted as an environmental impact statement was, by its own author's admission, "rather broad and cursory." The ecologist George D. Davis, a former staff member of both the Temporary Study Commission and the Adirondack Park Agency, prepared the assessment under less-than-ideal conditions. He had about two months to complete it and had to work with the company's flimsy preliminary efforts. Davis had written to Paparazzo, "I found the plots had no relation to feasibility and thus were only a pretty picture. The most egregious example is the equestrian area on top of a mat of sphagnum moss over stagnant bog water!" Furthermore, the plans Ton-Da-Lay submitted to the DEC did not correspond to Davis's recommendations. He had advised against the impoundments planned in swamps, had recommended that the company set guidelines for the protection of the St. Regis Canoe Area, and had encouraged Paparazzo to avoid subdividing around Dry Channel Pond in order to protect the habitat of spruce grouse and loons.[38] The DEC outlined other deficiencies in the Ton-Da-Lay application, including coverage of only one thousand acres rather than the entire tract; inconsistencies between maps; inadequate attention to electric service and sewage disposal; and no consideration of the development's impact on local services. The incompleteness of Ton-Da-Lay's application was a point of contention throughout the hearings, as Paparazzo opposed the broad scope of the DEC's inquiry and refused to supply the department with every detail requested.[39]

Diamond expanded on Carine's recommendation to make a broader statement about the impact of Ton-Da-Lay and the state's role in protecting natural resources. In his decision, Diamond quoted at length DEC witness Jerome W. Jensen, who had testified that the second-home subdivision would have detrimental effects on the adjacent Forest Preserve. His studies indicated that in order for public lands near Ton-Da-Lay to maintain their "wilderness character," only 350 to 500 persons should use them at one time. Building a community for up to twenty thousand people on a tract bordering state forestlands would dramatically increase pressure on the recreation-centered environment. Jensen noted: "any view of Dry Channel Pond that included obvious man-made development would have a strong effect on the mental appreciation of wilderness, hence degrading the experience and value."[40] Wilderness was more than a place; it was an experience involving seclusion and immersion in an area free from "man-made development." Restricting development on private lands based on how it affected "mental appreciation of wilderness" went far beyond pollution concerns. Here was the application of the Adirondack Park's recreational character to its ultimate extent, taking into account the psychological effects of development. Whereas most campers had, from the 1920s through the 1950s, embraced the comfort and security that came with elaborate recreational facilities, now peace of mind in the wild required significant distance from amenities more commonly found in cities and suburbs. A place had to look like wilderness, which, particularly after the federal Wilderness Act of 1964 and the Adirondack Park Agency's land plans, left little room for large numbers of people, let alone a neighboring second-home enclave.

Diamond did not stop with mental considerations. He also applied the APA's land-classification system that went into effect the day he issued his decision. Diamond argued, "The environmental impact of this development cannot be treated in a vacuum disregarding the Adirondack Park Land Use and Development Plan," and concluded: "The Applicant's plans . . . would adversely affect the health, safety and welfare of the people of the State and the natural resources thereof."[41] He justified his decision on both technical and principled grounds. Regarding the former, he made the case that Ton-Da-Lay's plans were inadequate, that the applicant did not prove the safety and healthfulness of its water-supply system. More ambitiously, Diamond affirmed the DEC's power to determine whether the project was in the public interest—that is, whether it benefited people and protected natural resources.

A furious Paparazzo dismissed the decision as a "sham" and took his case to the courts. In October 1973, his lawyers filed suit against Diamond, contending that the DEC had no statutory authority to reject the Ton-Da-Lay project on environmental grounds.[42] Essentially, the Ton-Da-Lay camp carried into the courtroom its earlier objections to the scope of the DEC hearings. The appellate division of the state supreme court rendered its decision on May 16, 1974. Even though the court upheld the DEC's rejection of the water-supply application, in a peculiar twist, the Ton-Da-Lay petitioners celebrated and the state respondents appealed.

The court validated the DEC's decision on narrow technical grounds, agreeing that Ton-Da-Lay's plans would not ensure proper protection of the water supply. But the court dismissed Diamond's broader justification on environmental principles: "We cannot permit such an extensive and unqualified finding to stand as a true measurement of the 'public necessity' aspect of this proceeding." The presiding justices agreed with Ton-Da-Lay's attorneys that Diamond had overstepped his statutory power, and they instructed the DEC to work with Ton-Da-Lay so that the company could correct the deficiencies in its plans and come up with a satisfactory solution.[43] All would not be smooth sailing for Paparazzo, however, once the APA stepped in and flexed its muscle.

ENVIRONMENTAL PROTECTION OR TYRANNY? THE APA VS. HORIZON AND TON-DA-LAY

Before the Adirondack Park Agency and Ton-Da-Lay began their protracted fight, the agency received the unpleasant, but not unexpected, news that other developers were set to challenge its legitimacy in court. Among those challengers was the Horizon Corporation, which in December 1974 alleged that the state, by approving the private land plan, "without compensation appropriated an interest" in its property. The corporation's lands in St. Lawrence County were now classified as rural use and resource management areas, limiting the total number of structures that could be built there to 1,608, far fewer than the 6,955 planned. Horizon sued for $36 million, but ultimately would have to walk away without the money it sought from the state.[44]

On October 21, 1976, the New York State Court of Claims dismissed Horizon's claim. Judge Henry W. Lengyel explained, "Esthetic, open space, and environmental considerations are valid bases for regulation in the Adirondack context.... The Adirondack Park is a resource of greater than local concern, and has been so declared by the Legislature."[45] Albany had a long record of singling out the Adirondack region for special protection, and that history was significant in justifying the continued exercise of state authority in the park. Though Horizon did not challenge the constitutionality of the APA, the state won a ringing endorsement of its power to manage the park's resources—one that would figure in future court decisions. Horizon's limited ambition left Ton-Da-Lay with the task of attempting to abolish the agency.

Paparazzo began making preparations for the subdivision of a portion of his property along the Jordan River to be called Jordan River Estates. In an attempt to conform to the new APA guidelines, he planned a much smaller development than the one the DEC had rejected: thirty ten-acre lots on fourteen hundred acres, compared to the earlier plan for 301 units on one thousand acres. A lot owners' association would enforce protective covenants similar to the ones proposed with the first development plan.[46] The APA caught wind of the steps Ton-Da-Lay was taking in spring 1974 and exchanged a series of increasingly contentious letters with Paparazzo and his attorneys.

In response to the executive director Richard A. Persico, Paparazzo asserted that the court's decision in *Ton-Da-Lay v. Diamond* was a rebuke against the APA's authority: "that the Agency wishes to review the *entire* plans . . . is not within the scope of the powers of the Agency as we read the statute. . . . I trust that the [APA] is not attempting to establish this precedent . . . since this corporation as a taxpayer in this state is not interested in assisting to defray any such judgements [*sic*]."[47] Confident in their rights as a tax-paying corporation to use their property as they saw fit, the company's representatives continued surveying the land and marking lots, and sought and received the approvals of the state health department, Altamont, and Franklin County.[48]

Undeterred, the APA continued to assert its jurisdiction and urge Paparazzo to cooperate. Not persuaded by Paparazzo's interpretation of the APA act, in spring 1975 Persico dispatched a project review specialist to investigate. On the Ton-Da-Lay property, he observed wetlands along the Jordan River and the distinctive flora and fauna they hosted. According to the specialist, these natural features qualified the area for resource management classification, and thus the proposed second-home community was subject to APA oversight. Still, the intransigent company refused to recognize the APA's authority.[49]

On August 1, 1975, attorney general Louis Lefkowitz sought an injunction to prevent the marking and sale of lots on Jordan River Estates until the APA had a chance to review the project. The complaint went further and insisted that Ton-Da-Lay "fully and completely remove . . . any stakes, survey markers, signs or other physical markings of any kind . . . and to restore the land within the Jordan River Estates subdivision to its natural condition." Here the APA revealed its perspective on the land: the environment's "natural condition" should be free from any signs of overt human manipulation. On August 5, a temporary restraining order was issued, halting work on Jordan River Estates.[50]

In response, Paparazzo challenged not only the injunction, but also the APA's legitimacy. He charged in a sworn affidavit, "The Agency has unreasonably denied permits to applicants, and has so impaired the power of disposition of land within the Adirondack Park as to render the legislation unconstitutional and invalid." Paparazzo pointed to other developers that had suffered at the hands of the APA, including Horizon. The guidelines' ambiguity, Paparazzo contended, prevented developers from meeting requirements and kept towns from formulating their own zoning ordinances. Worse, he asserted, the private land plan was no less than the state's attempt to confiscate property. Ton-Da-Lay had invested more than $3 million in the tract, but after passage of the APA law Paparazzo claimed that no prospective buyer would offer more than $100 an acre for land where development was restricted. If he sold his more than eighteen thousand acres at $100 an acre, he would suffer a loss of $1.5 million. He alleged that the state had ulterior motives in limiting land use: "the legislation is for the additional purpose of reducing land values to such a point that the individual land owner will have no choice but to sell his land to the State of New York, for inclusion in the Adirondack Park . . . because of the inability of the land owner to use the land for any purpose."[51]

New Yorkers watched the case anxiously. As a state legislative task force mulled over changes to the APA legislation, the Adirondack Park Local Government Review Board railed against the agency, North Country municipalities called for its abolition, protestors organized a march in Albany, and Adirondackers drove around with bumper stickers that read "Adirondack Park Agency—Another Word for Tyranny." APA opponents rallied around Ton-Da-Lay in what one Adirondack resident called the "Battle of the North Country."[52] The Franklin County legislature resolved to intervene on the developer's behalf and urged other municipalities to join them in their attempt to deal the APA "its death blow." During the following month, Justice Guy A. Graves allowed eighteen parties to intervene in the case. Joining Franklin County on Ton-Da-Lay's side as either interveners or *amici curiae* were, among many others, Altamont, Harrietstown, and the village of Tupper Lake. The APA, for its part, received assistance from the Kildare Club, Adirondack Council, Adirondack Mountain Club, Sierra Club, and Association for the Preservation of the Adirondacks. In December 1975, Ton-Da-Lay's attorneys asked Graves to lift the injunction, but it remained in place until May 1976, when oral arguments were set to begin.[53]

Much was riding on the outcome of the case. Adirondack towns and counties sought to preserve their power over land-use decisions. Jan Plumadore Jr., counsel for the restive municipalities intervening on Ton-Da-Lay's behalf, explained to Justice Graves that the people he represented were "on the verge of armed violence" in response to the state's assumption of greater power in their communities. On the opposing side, the interests of environmental groups and the owners of property adjoining Ton-Da-Lay dovetailed in their wish to prevent environmental degradation. Ruth and Ralph Friedman of the Kildare Club asserted their "vital interest" in preventing "Uncontrolled development [that] could result in erosion and pollution of streams and water courses which would affect our own property."[54] Justice Graves decided that questions of law and fact in the real estate developer's challenge to the APA's constitutionality would have to go to trial. He did lift the injunction, though, and allowed Ton-Da-Lay to continue preparations for sale.[55]

Meanwhile, things took a turn for the worse when two young Adirondackers attempted to start a bonfire in protest against the APA. In October 1976, APA counsel Robert C. Glennon stumbled upon twenty-five-year-old Brian Gale and nineteen-year-old Jonathan F. Barnes, both of Tupper Lake village, covering the agency's Ray Brook office in gasoline. Glennon subdued the two men and the state police charged them with attempted arson and assault.[56] Audrey Casier, who, in March 1976, had cofounded with her husband and other embittered land developers the anti-APA Adirondack Defense League, accused the agency of "provoking [the] violence" committed by Gale and Barnes. She explained, "Frustrated minorities burned cities in a struggle for freedom and civil rights. . . . I implore the Legislature and population at large to intervene before the continuing provocations by the Adirondack Park Agency result in a real conflagration."[57] Members of the defense league fancied themselves modern-day revolutionaries in common cause with civil rights activists, and Casier attributed APA opponents' crimes and intemperate language to the state's unjust policies.

Some Adirondackers had little patience for such confrontational tactics and inflammatory rhetoric, even if they were inclined to agree with the Casiers' position. Dave Landolfe of Lake Placid wrote his local newspaper to express his disapproval for the Casiers and their fellow defenders. He characterized them as a "group whose own selfish and narrow motives are clumsily hidden behind supposedly pure motives, self sacrifice, patriotism and service to the local people." Landolfe added, "Many non-wealthy people would not be able to afford to buy land and others couldn't afford to hold on to it because of taxes. . . . Historically, the poor people have always been pushed off the land with development." Whatever the substance of the arguments, Landolfe was disconcerted by the tone of the debate: "I personally know of many people who don't agree with the tactics of this group but who are intimidated by the atmosphere of violence and mob tactics. . . . Is this the freedom and self-respect this group keeps telling us about?" Landolfe's plea for civility failed to calm tensions.[58]

The changes that took place between the early twentieth century, when Franklin County became the center of resistance against state conservation laws, and the 1970s, complicated a fundamental question at the heart of the debate over the Adirondack Park Agency: what really was best for the average Adirondacker? During the conflict surrounding the creation of the Adirondack Park, residents, motivated by their moral ecology, defied control by both private and public interests, which seemed to work in concert to suppress locals' rights as stewards of the land. Hunting, gathering, and timber removal were crucial to residents' day-to-day survival, and the restriction of their subsistence and small-scale commercial uses favored the recreational uses of wealthy outsiders. During the 1970s, by contrast, Adirondackers marshaled the principles of independence and self-reliance in defense of corporate land developers that set out to promote tourism and second-home ownership—the very same recreational development responsible for residents' persistent economic insecurity and nonresidents' significant sway over land-use policy. Moreover, real-estate firms proposed changes to the land far more transformative than the uses residents had defended decades earlier. Whereas around the turn of the twentieth century Franklin County residents fought to protect established resource-use practices threatened by state conservation laws, decades later they resisted new policies that were intended to maintain established uses at the expense of future development. Paparazzo was able to present himself as David in a populist struggle against the state Goliath by capitalizing on the genuine desire of Adirondackers for an economic boost, residents' long-held wariness toward the state, local lawmakers' wish to maintain power, and, more broadly, a political climate imbued with a populist, antistate conservatism that would send its standard-bearer, Ronald Reagan, to the White House in 1981.

The virtually unrestricted development on private lands and the state's dedication to extensive recreational development on public lands through the first several decades of the Adirondack Park's existence had failed to address residents' economic hardship. Proposals for large-scale second-home development left many wondering about the consequences for those unable to afford a vacation home. The opposing populist argument asserted that such vacation communities would raise property values and taxes out of reach of the working class and poor. Developers like Nelson, Paparazzo, and the Casiers would see their bank accounts grow, but what about ordinary Adirondackers? Landolfe's fatalistic sense that nobody had the interests of the poor and working class at heart, and that they would continue to struggle no matter the outcome of the APA battles, was well earned. By the 1970s, the tourist economy had enjoyed at least fifty years to work its much-ballyhooed magic, and still Adirondackers groped for an economic lifeline. Regardless of whether the APA or Ton-Da-Lay won their legal bouts, ordinary park residents would be subservient to either the state or to large corporations. If history served as a guide, most Adirondackers would be left out in the cold.

With passions boiling over, Adirondack residents closely watched the case of *Adirondack Park Agency v. Ton-Da-Lay*. Paparazzo's attorneys, Jeremiah M. Hayes and Adam R. Palmer, argued that overly restrictive land-use regulations constituted illegal taking of property. Second-home construction, they contended, was the most "reasonable" purpose for the land because it was the most profitable; and the APA law imposed "a total freeze on all private development."[59] Whereas many environmentalists' vision of wilderness was restricted to seemingly pristine green space, development advocates' conception of progress was limited to large-scale productive uses like massive residential development. Though the private land plan left room for alternative uses, developers did not want to consider potentially less profitable plans. Hayes and Palmer claimed that the APA law amounted to "spot zoning": "the singling out [of] a small parcel of land for a use classification totally different from that of the surrounding." Hayes did not stop there; he went so far as to assert: "It is not a function of the [state's] police power to control the use of private land for the benefit of critical plant of wildlife habitat, scenic vistas, recreation, or to demand that only commercial forestry or agriculture be practiced."[60] Here was a Hail Mary of an argument whose implications could have been extraordinary. It seemed as if Ton-Da-Lay's attorneys were challenging not only the APA act, but also *all* state environmental regulation of private property. Would the court affirm this argument and bring to an end an Adirondack Park that embraced both public and private lands?

The attorney general's office offered its rebuttal. According to the state's brief, the "Act does not seek to prevent development but merely seeks to regulate the nature and extent of development," for in no privately owned area was development shut off completely. Though the kind of vacation community Paparazzo had originally proposed would violate the guidelines of the private land plan, less dense development supplemented by other uses, such as sports hunting and controlled logging, were permissible. Moreover, the legislature had approved the plan's classification system and the permissible uses enumerated therein. The state argued that it was incumbent upon the courts to defer to lawmakers' judgment. Attorney General Lefkowitz also rebutted the spot-zoning claim:

> the fatal flaw of defendants' premise [of spot zoning] is that there just does not exist any "other similar area of land of a comparable size in the State of New York" (as alleged in the counterclaim), since the Adirondack Park region of the State is, truly, sui generis. This was authoritatively determined by the Legislature in the very first sentence of the "Statement of legislative findings and purposes" (801 of the Act) wherein it declared that "The Adirondack Park is abundant in natural resources and open space unique to New York and the eastern United States."

Thus, it was not the Ton-Da-Lay tract that had been singled out, but rather the *entire* Adirondack Park. The state's long record of exercising special regulatory powers to conserve Adirondack land and resources justified the continuation of that practice with the APA.[61] The two sides argued their cases on December 15, 1977, and on February 10, 1978, the court found in favor of the state. In upholding the APA's authority, the court affirmed the necessity of the Adirondack Park's unique protections for the benefit of all New Yorkers.[62]

Once the court came down decidedly on the APA's side, Paparazzo scrambled to devise a program that was both profitable and compatible with the Park Private Land Use and Management Plan. He proposed, in addition to smaller residential development, a system for leasing out lands for recreation, a small lumbering operation, agricultural programs, a fish hatchery, the harvesting of sphagnum moss and peat, gravel mining, and Christmas tree farming.[63] His attempts to work within the confines of the APA act revealed how significantly his and other developers' vision of appropriate land use differed from the state's. If we were to imagine a phantom landscape that included Ton-Da-Lay and Horizon superimposed on top of parklands, we would see more than eleven thousand homes on formerly undeveloped lands; ski centers, golf courses, and other recreational features to compete with other private and state-run facilities; and almost eighty thousand people flooding into sparsely populated towns during the summer, only to return to their homes once outdoor temperatures dropped. The effects on people and the land would have been hard to predict, though undoubtedly they would have been dramatic and enduring.

CONCLUSION

On January 6, 1981, a nondescript announcement in the "Real Estate Transfers" section of the *Adirondack Daily Enterprise* announced a transaction between Ton-Da-Lay, Ltd., and Geoffrey A. Cobham of London, England. The notice was an obituary of sorts, announcing the death of Louis Paparazzo's dream for Ton-Da-Lay. Unsatisfied with the income earned from his property and his dealings with the state, in December 1980 Paparazzo sold 64 percent of his tract, a total of 11,740 acres, to Cobham for $1.265 million.[64] Paparazzo's nightmare at Altamont was over, but the wounds of his battles with the state remained raw in the Adirondacks.

During the controversy over second-home development in the 1970s, new environmental agencies asserted their power to regulate land use in the Adirondack Park, and they received support from courts that affirmed the supremacy of the state's authority over the region's counties, towns, and villages. Albany's history of protecting Adirondack resources for the benefit of all New York residents subjected major developments to intense scrutiny. "Forever wild" may have applied only to the Forest Preserve, but the state determined that the health of public lands was contingent upon the integrity of private lands. Backed mainly by middle-class New Yorkers, second-home owners, and environmental activists, the DEC and APA enforced restrictions rigorously during their early years, and they exploited the flaws and incompleteness of proposed developments to prevent large-scale subdivision of private lands. Though administrators and judges made important decisions in this story, ordinary people were involved every step of the way. As the fate of the Adirondack Park became a local, regional, state, and national concern, they formed advocacy groups, lobbied Albany and local governments, participated in hearings, wrote letters, drew up and signed petitions, organized protests, and raised money to cover court costs. The grassroots conflict was far more complex than surface-level analysis would suggest.

Although Horizon and Ton-Da-Lay met similar fates, the differences in their stories do much to illuminate unacknowledged truths about the conflicts over land use in the Adirondack Park. Horizon met intense opposition from people living inside and just outside the park, yet Ton-Da-Lay won significant support. Why was the response so different in these two cases? First, the two companies were dissimilar. Horizon had a reputation in the American West and Southwest for being insensitive to the lands it subdivided. Even the most committed development advocate recognized the need to preserve the scenic and rugged nature integral to the Adirondacks' tourist economy. Furthermore, just as many Adirondackers were suspicious of outsiders in Albany who managed the park and city folk who visited the region's attractions, they also harbored animosity toward outside capital. Horizon lived up to its negative reputation when, for instance, one of its contractors incurred fines for cutting down trees in the Forest Preserve. On the other hand, Paparazzo promised ecologically sensitive development of less than 10 percent of his tract. He also bought a home in Saranac Lake, which helped him avoid the label of outsider.

Moreover, St. Lawrence and Franklin counties were different in many ways. Since about half of St. Lawrence County lay outside the blue line and the Adirondack Park embraced most of Franklin County, perhaps residents of the former county did not feel hemmed in by state regulations, and as a consequence, they felt less animus toward Albany. On the other hand, Franklin County residents drew on a long tradition of antistate resistance in their opposition to the DEC and APA. In addition, the universities in St. Lawrence County, particularly Potsdam State and St. Lawrence University, served as centers of anti-Horizon sentiment. Development opponents tended to enjoy the benefits of a middle-class income and leisure time that enabled them to recreate in Adirondack parklands. Franklin County, by contrast, was among the poorest counties in New York, and the promise of jobs was music to locals' ears. Average weekly income of permanent Franklin County residents amounted to $91.14 in 1968, compared with St. Lawrence County's $127.25. As a result, Franklin County's population lost more than eight hundred people between 1960 and 1970, while St. Lawrence County's population grew modestly by 752 residents over that same period.[65] Differing opinions in and around the Adirondack Park belied the notion that the APA conflict essentially pitted Adirondackers against "outsiders"; and the environmental, economic, and class-based arguments employed by both sides forces us to rethink the idea that the debate can be boiled down to environmentalists versus developers.

As we will see in the next chapter, conflict over the Adirondack Park Agency and land-use practices was not limited to second-home development. The 1980 Olympics in Lake Placid brought short-term and long-term changes that also tested the APA. Though it emerged from multiple court challenges punch-drunk but still standing, by 1980 the APA was under attack not only from its antistate adversaries, but also from environmentalists unhappy with what they believed were environmentally damaging projects approved by the agency. The APA's flexible classification system became the subject of much controversy, as the state and federal government expanded the Adirondack Park's recreational facilities so that the Lake Placid area could play Olympic host once again.

CHAPTER 7

OLYMPIC TRANSFORMATIONS, PART II

The 1980 Winter Games in Lake Placid

In November 1976, the Adirondack Park Agency (APA), its future threatened by a series of lawsuits brought by real estate developers, began hearings on the potential environmental impact of the 1980 Winter Olympics to be held in Lake Placid. Among the witnesses to testify at the first hearing on November 19 was Jim Frenette, a councilman in Altamont, a town that had joined Ton-Da-Lay in its suit challenging the APA's constitutionality. Frenette was there to show support for construction of seventy- and ninety-meter ski jumps at Intervale, the site of the 1932 ski-jumping competitions. According to Frenette, "in . . . a long-established and proven winter sports center, certain facilities are necessary. They become part of the surroundings and do not detract from the natural beauty of the area. People expect to see them and more important, people accept them as *a natural and integral part of the surroundings* and, of economic interest."[1] As advocates did with all recreational-development projects proposed for the Adirondack Park since the First World War, Frenette highlighted the positive economic impact new winter-sports facilities would have on the region. Yet the councilman also made explicit what was often an unspoken assumption of both development proponents and environmental activists: that recreational facilities, no matter how modern and elaborate, became, over time, accepted features of the landscape alongside mountains, waterways, and trees. In a modern, domesticated setting, human-made structures became *natural*.

Despite Frenette's positive perspective on development, not everyone expected or wanted to see elaborate structures in the High Peaks area so popular among recreation seekers. Rather than dispelling the fears of those who cherished the Adirondacks' wild and scenic qualities, the councilman articulated precisely what had many environmental activists up in arms over the jumps. Mountain roads, interstate highways, modern campsites, and ski centers had become accepted parts of Adirondack parklands. If the APA approved the Intervale jumps, they, too, would seem natural in a recreational landscape where human-built structures often overshadowed natural features. Even though shorter jumps already existed at Intervale and the APA's classification of the area as a hamlet had opened up the land to further development, according to nature enthusiasts, the ninety-meter jump would diminish the scenic value that drew thousands of campers, hikers, and mountain climbers to the High Peaks. In this heated contest, as in the parallel dispute over second-home construction, environmentalists emphasized the interconnectedness of Adirondack parklands. Building the jump in a hamlet area, they argued, affected the recreational uses of neighboring "forever wild" lands and sacrificed the park's essential character as a recreation-centered wilderness retreat.

Still, both sides in the debate shared a common point of view: all were concerned primarily with human uses over other considerations—even, at times, the land's ecological health. Advocates, including most residents of Lake Placid, for whom winter-sports facilities were an accepted part of their economy and their day-to-day lives, supported potentially lucrative spectator sports and sightseeing from a shelter atop the ninety-meter ski-jump tower. Opponents, on the other hand, privileged mountain climbing and nature appreciation in a wild setting. Neither side was unified, however. Some environmental groups favored the Olympics and the economic development it would bring, while even some of the most committed Olympic boosters feared that

Lake Placid would lose its character as a picturesque winter resort. Determining which projects were appropriate to Lake Placid and its environs proved to be a continuing source of strife, as opposing camps struggled to define precisely the appropriate uses and meaning of the Adirondack Park.

The debate's untidiness lay in the complexity inherent in the contested and incomplete modernizing process taking hold in the Adirondacks. During the Olympic Games from February 13–24, 1980, Lake Placid, a village of 2,731 year-round inhabitants, would play host to some fifty thousand guests a day (compared to about seven thousand guests a day during the 1932 Games). Whereas in many battles of the North Country nature enthusiasts and development advocates tussled over the meaning of wilderness, in the Olympic debate they clashed over what it meant to be modernized. Such a massive event raised vexing questions about the long-term impact of attempting to simulate a small city. Would Olympic boosters' promises come true, and underemployed and unemployed Adirondackers be able to find work? Would sports facilities built and improved for the games provide a consistent source of revenue for local businesspeople? On the other hand, would the area's natural beauty be permanently degraded? Would the village lose its rustic qualities? All of these questions arose from the ambivalent consequences of recreational development. The 1980 Olympics highlighted the growing pains that came with the Adirondack Park's transition into a modern wilderness playground.

"A MODERN SURVIVAL TECHNIQUE": TOWARD THE XIII WINTER OLYMPIC GAMES IN LAKE PLACID

Lake Placid officials and community leaders had been eyeing another Olympics for many years prior to winning the bid for the 1980 games. Promoters considered another Olympics necessary to revive the resort industry in Essex County, which consistently ranked among the poorest New York counties. As the Lake Placid area played host to a number of national and international sporting competitions in the decades after World War II, including the 1961 World Bobsled Championships and the 1972 World University Games, the common cold gave way to full-blown Olympic fever. Unsuccessful bids for the 1960, 1968, and 1972 Winter Games did not deter the government officials, local businesspeople, and prominent sportsmen who formed the Lake Placid Bid Committee and were determined to bring the prestige and the short- and long-term economic benefits of hosting back to the village. In fact, controversy surrounding the 1976 Winter Olympics unexpectedly opened the door to another Lake Placid bid.[2]

In May 1970, the International Olympic Committee (IOC) awarded the 1976 Winter Games to Denver, Colorado. The excitement surrounding Denver's victory did not last, however, as fears of out-of-control costs and significant environmental damage inspired determined grassroots resistance. In November 1972, almost 60 percent of Colorado voters rejected a referendum on the allocation of public funds for the Olympics, forcing the IOC to find a new host city. Lake Placid was among the contenders to replace Denver, but ultimately, Innsbruck, Austria, host of the 1964 Winter Olympics, beat out its competitors for the 1976 Games.[3]

The Denver debacle did not discourage Lake Placid officials in their effort to lure the 1980 Winter Games, even though it had burdened the bid committee with the IOC's skepticism about prospective American hosts and the fear that environmentalists might resist the Olympics to protect the Adirondack Park. As a result of Denver's withdrawal, the IOC stipulated that subsequent bidders must hold a local referendum to approve Olympic proposals. On October 16, 1973, the town of North Elba, which embraced the village of Lake Placid, held its referendum, and the results were 726 in favor and 576 opposed. Although the successful referendum did not indicate a groundswell of support in the town of about 5,000 residents, the bid committee pressed on. After securing pledges of spiritual and monetary support from Albany and Washington, D.C., Lake Placid's representatives traveled to Vienna, Austria, in October 1974 to make their case before the IOC. Once the bid committee celebrated Lake Placid's long history as a winter resort and promised a more humble event in contrast to what many believed to be the bloated, overpriced spectacles of recent years, the village's four competitors dropped out of the running and the Olympics were set to return to the Adirondack Park.[4]

Crucial to the success of Lake Placid's boosters in selling their village to the state, federal government, and IOC was their promise of an "Olympics in perspective" that would protect both the environment and taxpayers' wallets. The plans of both the APA and town of North Elba ensured that building for the Olympics would occur on lands that had been zoned for development. In addition, the presence of winter-sports facilities in and around Lake Placid, constructed for the 1932 Winter Games and as part of earlier state-development programs, meant that new projects would build on existing infrastructure rather than require the clearing of undeveloped land. Also thanks to the facilities already in place, Lake Placid's promoters promised a total cost of less than $50 million. According to the committee, federal and state funding ensured that North Elba residents would not see a local tax increase—an important promise since they had only finished paying off the debt from the III Winter Games in 1973.[5]

Based on the premise that construction costs had spiraled out of control in previous Olympics hosted by big cities, the Lake Placid contingent promised a more humble event that returned focus to the athlete. The 1972 Winter Games in Sapporo, Japan, epitomized the kind of Olympics Lake Placid hoped to avoid. In addition to fourteen new sports venues, the games brought to Sapporo, a city of one million people, "the consolidation of roads and highways, sewerage systems, transportation organizations, hotels and communications facilities." Included among the Olympic projects were two downtown subway lines and an underground shopping center. As a result, the Sapporo games cost some $1.25 billion to stage.[6] By contrast, Lake Placid's committee pledged, "We do not propose any large . . . public works projects . . . or multi-million dollar 'one time' sports facilities. The time has come to restore the Winter Olympics to their proper perspective, to take them out of the city and return them to the small, mountainous winter sports communities where they originated."[7] The Lake Placid Olympic Organizing Committee (LPOOC), which was composed of the former bid committee members, celebrated the Winter Games' return to the cozy atmosphere of their earliest hosts: Chamonix, France, St. Moritz, Switzerland, and, of course, Lake Placid—whose 1975 population was actually smaller than it was in 1932.[8]

Although Lake Placid's representatives were committed to returning the Olympics to their roots and to putting on a more modest event, advocates inside and outside the LPOOC made no secret of their desire to capitalize on the games and invest heavily in tourism and winter sports. Robert McEwen, the region's U.S. representative, called attention to the need for an economic boost in Essex County, which had been suffering and losing residents since the closing of a paper mill and a Republic Steel mine during the late 1960s. The loss of manufacturing and mining jobs left workers even more reliant on the seasonal tourism industry, resulting in an unemployment rate close to 18 percent in 1975. Alternative industries were hard to find in New York's North Country, where distance from major population centers, rugged terrain, and harsh winters limited options, and so village leaders doubled down on tourism and recreation.[9]

Once again, the Olympics, even when kept in "perspective," were a means to an end: long-term economic growth. The bid committee commissioned an economic impact study by Plattsburgh University's Technical Assistance Center, which estimated that visitors to Lake Placid would pump $31.9 million into the village and surrounding communities during the games. Over the decade following the Winter Games, new and improved winter-sports facilities, increased publicity, and special events were expected to bring in an additional $30 million in tourist dollars.[10]

Regardless of organizers' rhetoric, the XIII Olympic Games would have to be a more elaborate affair than the earlier Lake Placid Olympics. During the 1932 games, 252 Olympians competed in 14 events, while in 1980 more than 1,000 athletes, from 37 countries, would participate in 38 contests. In order to stage such a massive event, state and federal funding was needed for a new field house, a new ice-skating rink, expanded trails and facilities at Whiteface Mountain, a new luge run and cross-country trails at Mt. Van Hoevenberg, improvements to the Olympic arena, and new ski jumps at Intervale. Though not seeking urban renewal on the scale seen in Sapporo, Japan, Lake Placid's boosters promoted substantial development of recreational infrastructure. According to the J. Bernard Fell, the executive director of the LPOOC, "We are convinced without question that the lifeblood, the economic future of our community is dependent upon our keeping ourselves forever in the limelight."[11] With the 1980 Olympics, the LPOOC, New York State, and the federal government carried on in the Adirondacks the trend of recreational development that had begun in earnest after World War I.

Lake Placid residents generally favored—at least initially—the games' return to their village, and their support hinged on the expected economic benefits. A survey of village residents conducted by the environmental studies program and North Country research center at St. Lawrence University found "very solid support for the Olympics." The report's authors used blunt language in calling the Olympics "a modern survival technique for the village and for the fragile wilderness area which surrounds it." The majority of survey respondents agreed with this assessment. One eighteen-year resident wrote, "The updating of facilities, additional employment, and worldwide recognition of the United States and Lake Placid as prominent in the resort and sports world will be highly beneficial." Another local observed, "Lake Placid's industry is the resort industry. Anything that enhances the basic industry enhances the community."[12] Development continued to be the overriding concern among Adirondackers who were searching for ways to escape the economic doldrums—just as it had been in the push for campsite improvement, building on Whiteface Mountain, Northway construction, and the subdivision of lands for second-home communities.

Nonetheless, a number of Lake Placid residents expressed skepticism about the organizing committee's promise of an economic boom. The antistate antagonism so widespread in the Adirondacks did not blind these doubters to the potential abuses of power closer to home. One survey respondent asserted, "The promotion of the 1980 Olympics, far from a grassroots movement, is largely an effort to relive past glories, coupled with the desire for personal economic gain. [The organizing committee members] are local businessmen . . . who stand to benefit economically . . . whether the environment or the general population benefits or suffers." The Olympic organizers were indeed part of the village elite: elected officials, businesspeople, and prominent members of sports clubs. For example, Jack Wilkins, who earned an income as head of the LPOOC's marketing committee until he was fired for alleged nepotism, was a wealthy business owner and real estate broker who made no secret of his desire to turn a personal profit from the games. The skeptical survey respondent assumed that organizers like Wilkins were driven by self-interest rather than a concern the common good. A local stonemason, no doubt made cynical by years of false promises of economic revitalization, agreed that the benefits would not be shared: "I believe many jobs will go to outsiders on political basis and all people will be left with higher taxes, assessments and a lot of litter and trampled facilities." Lake Placid's eldest residents, many of whom had less-than-fond memories of the 1932 games, tended to be the most resistant to the Olympics. They worried that tax increases and inflation of food, land, rent, and heating-oil prices would stretch their fixed incomes.[13] These were the same arguments opponents were making against the Horizon and Ton-Da-Lay developments in the concurrent debate over vacation-home construction. During the 1970s, the hidden costs of development distressed many throughout the Adirondack Park.

Although most locals expressed little concern for potential environmental damage, that did not mean they disregarded the natural environment. Eighty-three percent of survey respondents noted that they did not belong to an environmental organization, but it would not be accurate to infer, as the report's authors did, that residents' attitudes "indicate a lack of environmental concern or awareness within the community." To be sure, many locals resented the environmentalist bogeyman; as one respondent said, "I feel that interference by outside groups of environmentalists . . . is in direct opposition to the good of the residents of Northern New York. . . . I question the elitist motives." Nonetheless, joining the Sierra Club was not a prerequisite for caring about the natural environment. Rather than disregarding environmental concerns, Lake Placid residents had faith in their own ability to protect the resources and scenery in their community without the involvement of nonresidents. A critic of the APA and environmental groups expressed his sense of moral ecology in this way: "Those of us who have lived here so many years are certainly not interested in seeing the natural beauty of the area ruined." Another resident echoed that sentiment: "I wouldn't support the Olympics if I thought it would" mar Lake Placid. One survey participant responded in terms familiar to the debate, boiling down the story to the environment versus jobs: "I would like to see the environment stay as it is, but I would rather see food for my family on the table."[14]

Even though Lake Placid residents tended to favor feeding their families over protecting the natural environment, many of them were apprehensive about the possibility that the Olympics would alter their community's character as a small resort village in the scenic High Peaks. Most residents' worries stemmed from their fears of the games' urbanizing effects, which had the potential to transform their home. In the short term, locals dreaded the increased traffic and noise levels that the tens of thousands of visitors would bring. In fact, a number of Lake Placid residents planned to leave their homes during the games to avoid the inevitable hassles. Florence O'Neill would be one Olympic refugee: "I just don't want to be here for the $2 hot dogs and the crowds and the traffic."[15]

Beyond the short-term headaches of playing Olympic host, many Lake Placid residents were concerned that the games would result in a lasting transformation of their home. A sizable minority was unequivocal in its desire to maintain the village's small, peaceful, and scenic atmosphere that had kept them there year-round. One survey respondent put her view succinctly: "Lake Placid is a small community, let's keep it that way." For this local, like other year-round residents and second-home owners, keeping her community small meant keeping outsiders away. In addition to the desire to be left alone, some feared that Olympic construction would destroy the natural features so crucial to tourism and recreation in the village, and thereby have a detrimental economic impact. Rather than having to choose between jobs and scenic nature, there existed the real possibility that Lake Placid residents would be able to enjoy neither. Edna Broquist, the president of the Lake Placid Business Association, said in an interview, "I think everybody is proud of this little town, and they don't want it to become . . . gaudy. . . . Because it's a beautiful country, and we don't want it to change." Shirley Seney, the president of the Lake Placid Central School Board of Education, had profound questions for the Olympic planners: "What is our community going to be after the Olympics? Is it going to be our town again? Are we going to get back to being everyday, ordinary people that are working hard to make a living[?]"[16] Whereas Broquist had moved to Lake Placid and Seney had been born there, both women cherished the attributes of the village that set it apart from big cities. At stake with Olympic development was the ability of residents to shape their village's character.

In their consternation over the possibility that their community would sacrifice its essential qualities, and in their desire to keep Lake Placid a small, rural village, many residents seemed to have more in common with the reviled environmentalists than they would have cared to admit. According to Theodore Hullar, the chairman of the Sierra Club's Atlantic chapter, the Olympics must not bring "exploitation by outsiders [which] has been the case over and over and over again, as natural and environmentally sound areas of our country have been exploited." R. Courtney Jones of the Adirondack Council, a consortium of groups committed to protecting the Adirondack Park's natural environment, asserted: "if the games can take place in an atmosphere where athletic competition is more important than commercial rivalry . . . then, we may achieve, for the first time in years, something close to an Olympiad than to a world's fair."[17] These words could have easily come out of the mouths of the Lake Placid residents who wanted to preserve their village's rustic attributes. Development and commercialization, most agreed, should only go so far in a small resort town. Even the LPOOC had promised to keep these Olympics in perspective. Still, despite the shared wishes and concerns of the organizers, many Lake Placid residents, and environmentalists, fault lines scarred the common ground they occupied. As the Olympic plans were closely scrutinized and gradually put into effect, those fault lines shook and created massive rifts between, and even within, competing factions.

"DAMN UGLY": THE SKI-JUMP CONTROVERSY

The most controversial and divisive proposal for the Olympics involved plans for a new ninety-meter ski jump at Intervale, which, once completed, would be the tallest structure between Albany and Montreal. As we have seen, observers often placed the Adirondack landscape on a spectrum between wild and developed—the former represented by pristine-looking green spaces and the latter represented by cities bursting at the seams with buildings and people. The meaning of "forever wild" was a source of contention throughout the history of the Adirondack Park—as seen in the contests over campsites, the Olympic bobsled run, Whiteface Mountain, the Northway, and second-home construction. Yet no less flexible was the definition of what it meant to be modern. Furthering the development process did not come with a license to destroy and build indiscriminately. After all, it was in early twentieth-century cities that planners first employed zoning to render the landscape more legible. The Adirondack Park Agency had adopted the urban planner's tool to ensure limited and orderly development in the Adirondack Park. However, the ski-jump controversy highlighted, for some environmentalists, the inadequacy of the language in the APA's land plan, which classified Intervale as a hamlet area where development should be concentrated. Rather than settling differences over land use, the APA act provided a new vocabulary in the verbal and legal fights over construction projects.

The competing factions' arguments came to light in press reports and in two sets of public hearings. In November 1976, the federal Department of Commerce's Economic Development Administration (EDA) held hearings on the draft environmental impact statement prepared by Sasaki Associates. From November through late December 1976, the APA held its own hearings on the proposed ski jumps. Although most projects in hamlet areas could proceed without APA review, the jumps came under agency scrutiny, as required by law, because they were more than forty feet tall. The hearings created a forum for a fierce public debate over wilderness aesthetics: if yelling were an Olympic sport, there would not have been enough medals to go around at the hearings. For the LPOOC, government officials, businesspeople, labor unions, and other supporters of the Intervale jumps, Olympic facilities were essential not only to the Winter Games but also to long-term recreational uses. Environmentalists, on the other hand, worried that the expanded jump complex would interfere with established recreational pursuits. Since fifteen-, twenty-five-, forty-five-, and seventy-meter jumps already stood at Intervale, debate centered not on the ecological consequences of construction but instead on the ninety-meter jump's visual impact. The jump tower would reach 266 feet into the sky, far above the tree line, and be visible from a number of vantage points in the High Peaks area so popular among mountain climbers and sightseers.

According to the LPOOC and supporters of the ski jumps, the Olympics were crucial to the economic improvement of the Lake Placid area, whose survival depended, to a significant degree, on winter sports. New infrastructure, they argued, was essential to building on the legacy of 1932. These would not be the spartan wood-and-steel jumps from the III Olympics. During the planning phase, architect Karl Martitsch used computers and wind tunnels to compute optimal ski-jumping conditions. Based on his calculations, Martitsch designed a moveable starting platform for the ninety-meter tower that could be adjusted based on wind velocity and weather conditions. A refrigerated slope and snowmaking machines further ensured a controlled jumping environment and inspired LPOOC president Ronald MacKenzie to remark, "We will have the first weather-proof Winter Games."[18] Encouraging profitable development required the exertion of a significant degree of control over the natural environment.

However, the taller jumps, unlike campsites, mountain roads, and ski centers, had little utility for mass recreation, as their use was limited to well-trained athletes. The jump complex, then, would function as a site of spectator sports, and the ninety-meter tower also would serve as a perch for sightseers. Just as travelers paid a toll to ascend Whiteface Mountain in automobiles and an elevator car, Intervale's visitors would pay a fee to take an elevator to the top of the tower and marvel at the beauty of the surrounding High Peaks. The jumps, like the reshaped Whiteface, were supposed to generate badly needed revenue for the village, town, and county. As the *Lake Placid News*, whose editors dismissed environmentalists' aesthetic concerns, argued, "You can't debate beauty on an empty stomach."[19]

Aesthetic concerns, according to ski-jump proponents, were a canard anyway, since the towers would be ornamental additions to the landscape. Designs called for concrete towers and steel slopes to extend outward from hills in a complex mingling of nonhuman nature and technology. The craftsmanship on display, according to J. Vernon Lamb Jr. of the LPOOC environmental committee, would accentuate nature's beauty: "It will be almost an art object." Ramon Lopez, the general project manager of Gilbane Building Company, the LPOOC's lead contractor, said of the jumps: "They are a monument. . . . As you drive into the area they tell you that you are in a winter sports facility."[20] For the LPOOC and its contractor, the ski jumps were improvements—additions to the park that gave it beauty and purpose. The jump's aesthetic value was based, in large part, on its usefulness.

Moreover, the LPOOC had pragmatic reasons for selecting Intervale instead of alternate sites considered in the environmental impact statement. Most obviously, it was close to the heart of Lake Placid, making travel between sports venues faster and easier. The Intervale environment, where wind and weather conditions were conducive to ski jumping, also made it a desirable site. In addition, the LPOOC argued, the environmental effects of construction on the land and waterscape would be negligible. Builders would work on a landscape already "disturbed" by the earlier construction of jumps, parking lots, and spectator grandstands, and so minimal clearing of earth and vegetation would be necessary. With North Elba set to take charge of the complex after the Olympics, the physical and administrative infrastructure was in place to ensure the ski jumps' profitable after-use.[21]

The APA had opened the door to further development at Intervale when the agency classified the site as a hamlet. "Hamlet areas," according to the APA, "will serve as the service and growth centers in the Adirondack Park. . . . In these areas, a wide variety of housing, commercial, recreational, social and professional needs of the Park's permanent, seasonal and transient populations will be met." The town of North Elba, for its part, had zoned the area as resort residence, which allowed for skiing sites. Lake Placid resident and ski-jump supporter Margaret Dewey took her place in an endless line of development advocates by pointing out: "Intervale is not wilderness."[22] In this instance, the APA and the state legislature had affirmed Dewey's position with the Adirondack land plan. In the same way that Whiteface had lost its distinction as wilderness with the construction of the mountain highway, Intervale's wilderness status seemed to disappear in many people's minds once ski jumps had been built there in the late 1920s.

Since the conflict over the ninety-meter ski jump revolved around the tower's aesthetic impact, Sasaki Associates conducted an in-depth examination of its effect on sightseers' views. The company looked at the distant view (between 2.6 and 8.2 miles away), the middle view (between 1 mile and 2.3 miles away), and the near view (extending 0.75 miles away from the tower). From the distant view, where climbers on such popular peaks as Sentinel and Algonquin could see the tower, "views of the proposed ski jump . . . are considered to be of minor significance since the ski jump will also be viewed in a large context of the urbanized area of the Village [of Lake Placid] and Town of North Elba. Furthermore, the ski jump from these distances will appear very small."[23] Noteworthy in this description was the consideration of the jump in its environmental context. The appropriateness of the jump tower's presence depended upon one's literal point of view. When viewed from a distance, the ninety-meter tower would have an "urbanized area" as its backdrop, and thus fit snugly into surrounding development. In other words, the jump could not ruin a scenic vista already disrupted by the village of Lake Placid. Past development had made the jump a natural part of the landscape. The implication was that the tower, if viewed against a more natural backdrop, would be an eyesore.

Indeed, Sasaki acknowledged that from the middle and near views the jump tower would diminish observers' appreciation of High Peaks' scenery. Most striking would be the tower's prominence vis-à-vis John Brown's Farm, a National Historic Site known for its rustic qualities. Although the tower would loom over state lands and cast a shadow on John Brown's Farm, Sasaki asserted that the jump tower was consistent with the very inconsistency of the surrounding landscape: "The open and developed nature of the airport and horse show ground areas and the mineral extraction area northeast of New York State Route 73 contrast greatly with the pastoral nature of John Brown's Farm and the valley corridor of the West Branch of the Ausable River."[24] According to champions of the Intervale site, the development pattern in the area reflected a mixture of uses, and the jump tower did not interfere with established practices. Completed ski jumps would make concrete once again the well-established trend that development begot more development.

Refuting environmentalists' contention that the ninety-meter jump would mar treasured scenic vistas involved a rhetorical legerdemain on the part of the jumps' defenders. Richard F. Galehouse of Sasaki Associates noted of the jumps: "If this object were a 26-story apartment building in slab of broad dimensions, it would have an adverse impact." He later added, "it is clearly the intent of the Adirondack land use master plan to concentrate urbanization and development activity in the hamlet areas." Jump proponents' position was not *anything goes*; rather, it was based on careful consideration of the village's character and the surrounding environment. Ski jumps were consistent with the visual scheme of a winter resort, but a tall apartment building was not. Within the spectrum from wilderness to metropolis, there existed many shades of wild and developed. But an exchange between DEC counsel Philip G. Gitlen and Galehouse muddied the latter's point. When Gitlen asked, "an urban landscape is one in which man's structures dominate the landscape?" Galehouse answered: "There are parts of the village which approach that, yes." He even went on to assert that the existing jumps were "quite an eyesore."[25] Clearly the public debate often raised more questions than it answered: if parts of Lake Placid resembled an urban landscape, why would a large apartment complex not be an appropriate addition? If the old jumps were eyesores, what would make the new towers aesthetically appealing? If the law was not always a reliable or welcome guide, what criteria determined which structures were appropriate in a given landscape?

A combination of natural features and established land-use practices seemed to determine which structures belonged in a particular landscape. For instance, jumps had no place on top of Whiteface Mountain, whose almost-mile-high stature would have resulted in a tall stack of dead ski jumpers. Yet past development had also made construction impractical in places where building jumps might be possible. As LPOOC attorney Robert J. Kafin said in frustration after being pressed about alternative jump sites: "You could build one on a flag pole, or in Yankee Stadium."[26] Carefully selecting which buildings to construct and deciding where to site them was essential even in the most developed cities so that, for instance, a factory was not built in Central Park. Indeed, John Lansing, a Lake Placid accountant, pointed out: "Central Park is an imposition of a natural environment in a man-made area."[27] The incongruities within an urban landscape that embraced both skyscrapers and massive green spaces did not make Central Park any less desirable. Therein lay the rub. As Altamont councilman Jim Frenette pointed out at the first APA hearing, any structure could become accepted as natural over time, and in the process facilitate additional developments as long as they could be justified within the rather expansive and flexible terms of established land-use practices.

Environmental groups brushed aside the LPOOC's arguments and pledged to protect the Adirondacks' scenery for the enjoyment of recreation seekers. To many environmentalists, the area was more than a land of imposing mountains, stoic trees, and meandering rivers, for these natural features constituted their ideal vision of a recreation-centered landscape. Whereas the ski jumps had symbolic weight for winter-sports enthusiasts, for many environmentalists the High Peaks represented the best of the Adirondack Park. Wilderness had two essential attributes, aesthetic and utilitarian. First, it had to *appear* undisturbed, as if only plants and wildlife flourished there. Second, it had to be *useful* as a site for leave-no-trace recreation. Without the concern for human uses, visual impact would have been irrelevant. If environmentalists wanted an untouched landscape, they would have called for the removal of the existing towers, and they would not have advocated for a different jump site within the Adirondack Park. The shorter jumps could stay because they did not interfere with recreation seekers' ability to appreciate the High Peaks' awe-inspiring vistas. Grace Pierce of the Wilderness Society, for her part, countered the argument that the jump tower would have a minimal visual impact: "This is like playing down the impact of a brown gnat in a bowl of gourmet cream soup."[28] High Peaks' scenery was akin to gourmet food consumed at upscale restaurants, something that nourished refined palettes. Unfortunately, Pierce's choice of words played into the stereotype of environmentalists as elitist snobs who sought to protect their own recreational pursuits above the economic interests of workaday people.

Elitism was often a fair charge against environmentalists, but that did not diminish their commitment to their cause. James Dumont, a Columbia University law student who represented the Sierra Club in the Adirondacks, may not have fallen into the same trap Pierce dove into, but his stridency made him few friends in the LPOOC and Lake Placid. Dumont argued that the 266-foot tower would have "the visual equivalent of putting an illuminated 26-story apartment building" at Intervale. After reviewing the draft environmental impact statement, Dumont concluded: "the beauty and isolated atmosphere of the High Peak area of the Adirondack Park may be lost forever if present Olympic plans go unchecked." If those words did not come out of the mouth of a brash young environmentalist, they might have resonated among the Lake Placid residents who feared losing their village's rusticity and beauty to overdevelopment. Nonetheless, Dumont's words elicited vigorous rebukes from the LPOOC, labor unions, and state and local government officials. Undeterred, Dumont pointed out the apparent irony that: "The state, at a cost of $100,000, is removing fire towers because they don't conform to the wilderness. The ski jump will be at least as visible from wilderness peaks as any fire tower."[29] This comparison was misleading, though, because the Department of Environmental Conservation (DEC) was removing structures from state lands designated wilderness by the APA, while Intervale was a hamlet. Dumont was not ignorant of the agency's classification; instead, the Sierra Club's priorities differed from those of the APA.

Thus, for some environmental groups, planners' argument that the existing ski jumps made Intervale the perfect spot for newer and taller jumps carried on a dangerous trend that had played out over and over again in the Adirondacks. Champions of development frequently pointed to existing structures as evidence that a particular corner of the Adirondack Park had lost its wilderness value, and environmentalists endeavored to halt the persistent pattern in which improvements led to more elaborate improvements. As Dumont noted, in areas the APA had classified as wilderness, the DEC had reversed course and had begun to remove "non-conforming" structures. However, recovering wilderness in places where thousands of people lived and infrastructure blended with the natural landscape would have been a quixotic task. The site's hamlet classification was irrelevant to Dumont and others because the ski jumps would have a detrimental effect on outdoor play in a treasured section of the Adirondack Park. Environmentalists fought battles against environmental degradation on several fronts; for at the same time that Lake Placid's plans for the Olympics lurched toward fruition, activists were in courtrooms defending the APA as protector of the Adirondack Park's natural resources. With Intervale, as with Ton-Da-Lay, environmental groups' overriding concern was the salvation of low-impact recreational uses enjoyed mainly by the middle and upper classes.

Even nature enthusiasts were divided over the ski jumps. The Adirondack Mountain Club, for instance, distanced itself from the Adirondack Council, which it deemed too aggressive in opposing the jumps and potentially endangering the Lake Placid Winter Games (just as environmentalists' opposition had once helped prevent a Denver Olympics). According to Edwin H. Ketchledge, "I feared the credibility of the Adirondack Mountain Club was impaired by our association with statements and policies contrary to the cooperation and moderation that has been our stand all along." He had climbed Algonquin, the closest High Peak to Intervale, 102 times, and asserted, "you cannot see Intervale and the ski jump unless you have binoculars. The alleged vision impact is an illusion without a solution." Ketchledge dismissed aesthetic concerns, and he and his cohort favored the economic and recreational benefits of the Olympics. Indeed, the group had a direct connection to the planning process. J. Vernon Lamb Jr., who had founded the Lake Placid chapter of the Adirondack Mountain Club, was chair of the LPOOC's environmental council. The Sierra Club, on the other hand, threatened to bolt the Adirondack Council for not being assertive enough in challenging the jumps, and resolved to participate in the APA hearings independently of the council.[30] The diverse and sometimes contradictory priorities of environmental groups created divisions within their ranks.

Even though environmentalists did not form a united front during the Olympic debate, critics often vilified the movement's adherents as uniformly radical and obstructionist. Olympic organizers and development advocates frequently responded to environmentalists as if the latter had been spawned from a cocoon during the 1960s. While many in the green backlash continued to stigmatize environmentalists as wealthy elitists, the stereotypes of activists broadened to include the filthy hippie. LPOOC vice-president Art Devlin relished recounting a confrontation with a development opponent: "He had glasses so thick with dirt he could hardly see through them, shoes all broken, gravy in his beard. He came up to me and started talking about the ski jump, and I said to him, 'I don't know about environmental stuff but you sure look like sight pollution to me.'" There *was* something different about the movement of the 1960s and 1970s, when it drew a broader, largely middle-class base. Perhaps in part as a result of their youth and physical appearance, this new crop of nature enthusiasts seemed more radical to many Adirondackers. As Robert Allen, the manager of the North Elba Park District and planning and engineer chairman for the LPOOC, commented, "I am a conservationist, I think that the environmentalists have bought themselves an awful dirty name by their opposition to all kinds of advancement and improvement. If you want to read by candle that is the true environmentalist."[31]

Allen would have been surprised to learn, however, that throughout the twentieth century disagreement over the nature and scale of recreational development in the Adirondacks was as constant as the presence of Whiteface Mountain. In fact, his unflattering characterization of environmentalists had been used to describe early conservationists. For instance, in 1930 Godfrey Dewey had called the Association for the Protection of the Adirondacks' challenge to the bobsled run "preposterous nonsense." Five years later, the public works commissioner Frederick Stuart Greene derided "professional conservationists" as "those who love nature from a swivel chair" for their opposition to Whiteface Highway—a project that had split self-described conservationists. Adirondackers were not a likeminded mass in their positions on recreational-development projects, and neither were nature enthusiasts. For its part, the Adirondack Mountain Club had, since its founding in 1922, blazed a twisting path in the various contests over land use in the Adirondack Park. The club had promoted both Whiteface ski centers, but then led opposition to the Hoffman Mountain amendment. And, while the group was championing the Olympics, it was simultaneously lending support to the APA's case against Ton-Da-Lay. Positions shifted as the political, economic, and physical terrain shifted. One consistent thread was the tension between advocates of mass recreation and those who desired to preserve the Adirondack Park's natural beauty for relatively low-impact recreational uses.

As evidence of some environmentalists' single-minded focus on scenic values and recreation, the Sierra Club and Adirondack Council pledged not to take legal action to hold up the ski jumps if they were built on an alternative site at Bassett Mountain, located about fifteen miles from Lake Placid, in the town of Jay. They were willing to sacrifice the Bassett site to save High Peaks' scenery for the benefit of mountain climbers and sightseers. Although ski jumps at Bassett would not rise above the mountains and, consequently, their visual impact would have been much less conspicuous than the structures at Intervale, a number of practical and environmental considerations made it a less desirable site. First, snow and wind conditions there were less favorable to ski jumping. More significantly, unlike the Intervale Hamlet area, the Bassett site was private land classified as resource management by the APA. As a result, the state would have had to purchase the tract, and builders would have had to construct a half-mile-long road, extend power and sewage lines to the complex, clear twenty-five acres of forestland, and excavate fifty-five thousand cubic yards of earth in a spot that the APA had deemed ecologically fragile.[32]

Ironically, the Sierra Club and Adirondack Council's focus on wilderness aesthetics moved them to take a position that would have resulted in the physical degradation of the Bassett site to protect the beauty of the High Peaks. Unlike the earlier Northway debate, in which route-C proponents lobbied to move the highway from "forever wild" lands to the more developed Champlain Valley, ski-jump opponents pushed to move the jumps from Intervale to the *less* developed Bassett Mountain. No wonder this peculiar position divided nature enthusiasts.

John Wargo of the APA called the ski-jump hearings "one of the first major test cases of the aesthetics effect in a wilderness area." Nevertheless, as we have seen, "mental appreciation of nature" was a significant rationale in the DEC commissioner Henry L. Diamond's rejection of Ton-Da-Lay's application in 1973. In fact, the DEC's Jerome W. Jensen, whom Diamond had quoted in his decision, testified at the Olympic hearings and called attention to the "psychological impact" the jump towers would have on visitors for whom "wild forest values are their primary objective." On the other hand, Jensen added, Lake Placid residents may "perceive this facility, and the Olympics, as important to them personally, important to them in their economy, and a sense of community enhancement."[33] Form and function complemented each other, as perception of the jump towers' aesthetic impact was contingent upon how one related to the surrounding natural and built environment. According to the Sierra Club and Adirondack Council, which seemed to privilege sightseeing over the ecological health of the Bassett area, mental appreciation of nature overrode concern for the environment's physical integrity. In the process, these environmental groups dismissed the very APA land-classification system they were attempting to save in the courts.

The Sierra Club and Adirondack Council would be at loggerheads with the APA once the agency gave conceptual approval to the Intervale jumps. On January 10, 1977, the APA's commissioners voted 6-4 (with one abstention) to allow the LPOOC to continue preparations for construction of the jumps. While acknowledging that the towers would have an adverse visual impact from short- and mid-range views, the majority of APA decision-makers argued that positive economic outcomes overrode aesthetic considerations. Essentially, they agreed that one should not argue beauty on an empty stomach: "We take official notice of the fact that the local economy, in an unfortunate condition at present, will be benefitted by the project and by the Olympics generally."[34] Though the Adirondackers and politicians who were resistant to state environmental policy often lumped together the APA and environmental groups, they did not form a united front against development. Indeed, the environmental movement itself split over the jumps, and the APA's approval of the project further alienated the Sierra Club and Adirondack Council.

The ski-jump debate was, in part, a struggle for the soul of the young Adirondack Park Agency. Of course, the LPOOC, real estate developers, and North Country politicians played an adversarial role against the APA—but so did environmental groups. The frustration with the state that both development advocates and environmentalists felt had been percolating for some time prior to the Olympic debate, as administrators' middle road seemed to disappoint everybody. Most distressing for environmentalists was what they considered to be the agency's spinelessness. The Sierra Club's Dumont said, "We hope that the actions of the Agency do not mark the beginning of a passive and weak role for it in evaluating major development projects in the Adirondacks."[35] According to APA chairman Robert Flacke (who later went on to head the DEC), the Sierra Club, Adirondack Council, and their allies on the APA staff sought to delay the review process and prevent the commissioners from rendering a decision on the jumps. The chairman, in his own words, "told the Environmental Movement to go jump in a lake." According to Flacke, the commissioners' approval of the jumps "was ... a major defeat for [environmentalists] ... If they had won it everybody would have been scared to death of them."[36]

Instead, the divided APA commissioners asserted their authority and the ski-jump opponents were left licking their wounds.

Flacke went further and charged, without convincing evidence, the Sierra Club and Adirondack Council with attempting to prevent the Olympics from happening at all.[37] Although the groups had drawn a line in the snow with the jump towers, they consistently expressed support for the Olympics. Moreover, they left in their quivers several arrows that could have punctured the Olympic balloon. They threatened to sue the state over the ski jumps, but instead secured through negotiations with the state assurances that the Department of Public Works would not widen highways in the Adirondack Park and that the DEC would devise plans to alleviate new pressures on wilderness areas. The APA, however, told environmentalists to go jump in a lake again, refusing to take part in any agreement. In April 1977, the commissioners gave unanimous final approval to the ski jumps with nary a peep from the Sierra Club and Adirondack Council.[38] Even with the APA roadblock in the rearview mirror, the road toward the Olympics was a slippery one filled with potholes.

Construction began soon after APA approval, but building the ski jumps proved to be a significant challenge. Human folly and the exigencies of capitalism turned out to be greater obstacles to the facility's completion than resistance from environmentalists. By November 1977, the two support towers had been erected, but in spring the following year, work halted when the contractor responsible for fabricating the structural steel went bankrupt. In an effort to ensure timely completion of the jumps, another firm, Gilbane Building Company, took the steel from the bankrupt firm's yard and transported it to Montreal, where the company Dominion Bridge completed fabrication and transported the jump back to Intervale. Meanwhile, erosion into the Ausable River during the building process also caused delays. Construction of both jumps, which ended up costing $6 million, more than double the original estimate, was complete by early 1979, in time for the Olympic pregames (see figure 11). Unfortunately, the pregames presaged the logistical problems that would later bedevil the Winter Games. The

FIGURE 11. Governor Hugh Carey speaking in front of the Olympic ski jump, photograph by Fred Ricard, 1981. Governor Carey speaks in front of the controversial ninety-meter ski jump built for the 1980 Olympics, in July 1981. Courtesy of the New York State Archives and Fred Ricard.

four thousand spectators leaving the jump complex after the warm-up competitions were tied up in a traffic snarl along Route 73 that was akin to New York City's rush hour. The drive between Intervale and Mt. Van Hoevenberg, which normally lasted five to ten minutes, took drivers an hour and a half. Since some fifty thousand people were expected to visit North Elba each day of the Olympic Games, one resident observed, "If this is indicative of how it's going to be in 1980, look out Charlie."[39] "Charlie" was not the only one who needed to look out.

As for the completed jumps, the verdict was split. Many continued to celebrate their beauty and potential to generate revenue for the Lake Placid community.[40] For others the jump complex represented the worst the Olympics could bring to a rural Adirondack village. Richard Persico, the executive director of the APA, later recalled the ski-jump debate: "there were those . . . who just did not want those towers to go up for reasons of visibility and now that I see them I can't help but agree that they had great wisdom and foresight. . . . They are damn ugly." Third-generation Lake Placid native Vera E. Littlejohn called the expanded complex "an aesthetic and environmental disaster." Warren Kane, a clerk of the U.S. Senate Subcommittee on Commerce Appropriations, which controlled the EDA's purse strings, lamented the commercialization of the games made manifest at Intervale: "I'm afraid you're gonna have these skiers going down off these things with more advertisements than an Indianapolis [race] car has. . . . It's going to look *like Broadway* going down the ski jump."[41] Though Kobot the Coca-Cola Robot, one of the many corporate mascots at the games, may have resented Kane's remarks, it was hard to argue that the Lake Placid Olympics, with its more than two hundred sponsors and suppliers, were highly commercial.[42] The games seemed to have brought yet another Broadway to the Adirondacks.

The image of Broadway proved to be a persistently popular urban metaphor for what was gained and lost with modern recreational development. Back in the 1930s, when an observer celebrated Whiteface highway as a mountain Broadway, the city had not yet taken on the grim cast it would during the urban crisis of the 1960s and 1970s.

By the middle of the latter decade, nobody involved in the Olympic debate wanted to associate with cities. The jumps may have represented economic improvement to some, but jump advocates insisted they were nothing like urban apartment buildings. To environmentalists, on the other hand, the jumps were too much like an apartment building, for they represented loss of the rusticity, natural beauty, and isolation that had defined North Elba's character. According to Warren Kane, the jumps were evidence that the crass commercialism associated with Broadway had taken over the town. A complex undertaking on an Olympic scale was bound to meet the expectations of detractors and champions alike. As the Winter Games approached and the effects of development became apparent, Olympic host's remorse began to set in.

"WE DIDN'T EXPECT THINGS TO HAPPEN THIS WAY": DEVELOPMENT AND ITS DISCONTENTS

Development in and around Lake Placid was not confined to the construction of sports facilities. A building surge accompanied preparations for the games. Sewage and power lines were extended to serve new businesses and homes. A new shopping plaza, two new subdivisions (Olympic Hills and Liberty Hills), and a new Hilton hotel joined the Olympics facilities as additions to the Lake Placid area. New shops also popped up on Main Street, and a number of stores were renovated. Peak employment of three thousand workers put more money in Adirondackers' pockets, and the publicity leading up to the games spread enthusiasm throughout Lake Placid. One local spoke on behalf of many village merchants: "I could show you a number of business places that were on the ropes. The Olympics bailed them out and there is no doubt about that."[43] Lake Placid businesses must have been in deep trouble, if it took the Winter Olympics, a global event hosted by the village twice in fifty years, to save them from collapse.

As building for the Winter Games proceeded apace from 1976 to 1979, environmental concerns mounted. Initial LPOOC plans called for the blazing of cross-country trails on wild forest lands near Mt. Van Hoevenberg, but resistance from environmentalists led the state to purchase new property so the trails would conform to the APA's master plan. On Whiteface Mountain, the DEC cleared trees for the addition of a new lift, which raised questions about the constitutionality of construction on "forever wild" lands. The DEC also angered conservationists when it widened ski trails on Whiteface to two hundred feet in order to conform to Olympic guidelines, even though the state constitution limited trail width to eighty feet. Ironically, back in the 1930s, well before the DEC and APA had come into existence, the Forest Preserve had been safer from Olympic development, thanks to the appeals court's ruling on the unconstitutionality of the bobsled run. In addition, the Sierra Club challenged the widening of Route 73, alleging that the state had reneged on its promise not to expand Adirondack roads, but its case was dismissed on the grounds that the Department of Public Works had been planning improvements to the road for some time. Over in Ray Brook, the EDA ordered a halt to construction of the Olympic Village so it could remedy a severe erosion problem. Though environmental organizations called on the state to address these issues and challenged the Route 73 repairs in court, they tended to play the role of paper tiger, threatening to sue but ultimately backing down. For example, the Adirondack Council delayed its suit over the unconstitutional Whiteface trails until after the games and ultimately declined to pursue legal action.[44] As a consequence, Olympic preparations proceeded with minimal interference from environmentalists after the ski-jump controversy had been settled.

Still, preparations did not go smoothly, as the organizers struggled to keep their promise of an Olympics in perspective. The games' cost, originally estimated at about $50 million, ballooned to $200 million. Inflation, construction delays, questionable building practices, evolving plans, bankrupt building companies, and 106 separate lawsuits brought by inadequately compensated contractors sent the price tag spiraling upward, forcing the LPOOC to appeal to Albany and Washington, D.C., for more money. As a result of the LPOOC's questionable management, the EDA made the release of additional funds contingent on the hiring of a new supervisor, Petr Spurney, who had overseen the 1974 World's Fair in Spokane, Washington.[45]

Problems would persist. As many had expected and feared, development brought soaring prices that squeezed many Lake Placid residents. Major Olympic sponsors such as Adidas and Nikon rented Main Street shops for as high as $65,000 a month. At the same time, ordinary visitors and village residents could not outspend the crown prince of Norway, who rented a house in the area for $55,000 a month. With dollar signs dancing in their heads, a number of landlords evicted their low-income tenants, many of whom resided in their homes without leases. For instance, Beverly Manning and her five children were forced to move from their Main Street home after the landlord, Jack Wilkins of the LPOOC, ignored her pleas to fix the nonfunctioning heating and water systems. He would not go ahead with the needed improvements until the Winter Games had brought bigger spenders to Lake Placid. "I was all ready to go see the Games," Manning told the *New York Times*, "but I don't really care a thing now.... This town can go fall in a hole, far as I care." Restaurants bumped up their prices as well. Diners at Main Street's Woodshed Restaurant, for example, would have to pay a $5 deposit to reserve a spot for each person in their party. Upon finishing their meals, they would then fork over $30 each on a reduced selection of entrees that had cost about $12.50 prior to the Olympics.[46]

To reiterate Shirley Seney's questions for the Olympics planners: "What is our community going to be after the Olympics? Is it going to be our town again?" Answers were uncertain before, during, and immediately after the games. Shop, home, and hotel construction seemed to cater to outsiders. Business owners increased prices to exploit an expanded consumer base, though sales turned out to be disappointing during the games.[47] Some Lake Placid residents lost their homes so landlords could rent to Olympic guests. Multinational corporations and large retailers sunk their hooks into the village. The fear among Lake Placid residents that they would have to sacrifice some of their village's essential characteristics proved to be well founded. Even Jack Shea, the North Elba town supervisor, speed-skating gold-medalist during the 1932 Lakes Placid Games, and one of the most active champions of the Olympics, lamented the XIII Winter Games' impact on his home. "We didn't expect things to happen this way," Shea told *Adirondack Life* a few months before the Games began. "This community has changed so much. It's not as homey a community as it used to be.... The Olympics have brought so much commercial interest. The beautiful open spaces on Main Street are gone."[48] Lake Placid residents went from experiencing the trials of living in a small village dependent on tourism to suffering through the growing pains of developing for the Olympics. Isolation, simplicity, rusticity, and scenic beauty—these were the qualities locals came to associate with their home, and they were at least temporarily sacrificed so the village could enter the world stage. Such was the lot of a modern wilderness playground.

CONCLUSION

After years of struggle, contestation, and controversy, in February 1980 Lake Placid hosted its second Olympics, and triumphs in athletic competitions began to overshadow the circuitous, uphill climb to the games. The XIII Winter Games have continued to occupy a prominent place in the national consciousness, thanks in large part to the U.S. hockey team's triumph over the Soviet Union at a time when tensions between the two nations were ramping up again. In Lake Placid and throughout the Adirondack region, nostalgia for the games has produced such headlines as: "Lake Placid: The Last Games of Their Kind"; "Through the Shadow of the Cold War, Lake Placid's Winter Olympics Went on"; and "It was the Best Time I Had in My Life."[49] Few remember the *Lake Placid News* headline from the opening day of the Games: "Wilmington Fears Health Hazard from Whiteface Garbage."[50] In addition to bringing athletes and guests from across the globe, greater fame, and new improvements to recreational infrastructure, the games also brought to the Lake Placid area tons of trash.

Though known mostly for the "Miracle on Ice" rather than "Garbage from the Mountain," the pressures of a much-expanded human presence in and around Lake Placid were a significant part of the Olympic story. The refuse piling up was not the only problem brought by the rapid development process. During the first few days of the games, a poorly organized busing system left many visitors stranded out in the cold for hours while the events for which they had acquired tickets went on without them. In response, Governor Hugh Carey declared a limited state of emergency and dispatched the Red Cross, school-bus drivers, and others to remedy the botched planning. The LPOOC's struggles led a German observer to comment, "The only amateurs here are the people running the games."[51] Though the haze of nostalgia may have obscured the contests waged over the Olympics' environmental impact, the temporary and permanent changes effected in preparation for the 1980 games again highlighted the costs of, and contentiousness surrounding, recreational development in the Adirondack Park.

As state planners had been doing for decades, the Olympic organizers expended copious amounts of money and resources to impose a modern style of land use on a largely rural and wild setting. Welcoming, housing, feeding, transporting, and entertaining great masses of people proved to be a tremendous challenge for the ill-equipped LPOOC. Despite prior development enacted by North Elba and the state, the incomplete modernizing process that had left Lake Placid a small, rustic community needed to be escalated for the XIII Winter Olympics. Pushing Lake Placid so rapidly along the spectrum from wilderness to city created new economic opportunities, but at the expense of the rustic characteristics embraced by residents and visitors alike. Beyond the bitter memories etched into the minds of Lake Placid residents who endured the hassles of simulating a small city, there remained marks on the land that carried on the ambivalent legacy of the Olympics. Skiers at Whiteface enjoyed new facilities and trails whose construction violated constitutional protections of "forever wild" lands. New buildings along Lake Placid's Main Street were evidence of economic promise, but their presence meant pedestrians would no longer be able to enjoy lost scenic views. While spectators cheered on ski jumpers at Intervale, visitors to nearby scenic and historic sites could not escape the shadow of the ninety-meter jump tower.

In retrospect, it is striking how similar the stated concerns and goals of most people involved in the Olympic story were. Though the rhetoric was heated, all participants agreed on the terms of the debate. Both champions of the Intervale ski jump and environmental groups favored policies that catered largely to recreation seekers from outside the region—drawn either by sports competitions or by High Peaks scenery. Environmentalists and development advocates both wanted to preserve the area's scenic beauty, which enriched sightseers' spirits and put money in local business owners' pockets. Nobody proposed turning Lake Placid into a Coney Island amusement park; in fact, everyone involved stated his or her commitment to preventing such an outcome. Despite Olympic promoters' characterization of environmentalists as obstructionists, their intervention would not have prevented development. In fact, they endeavored to relocate ski-jump construction, not stop it. All agreed that a tall apartment building did not belong in the Adirondacks, and all agreed that ski jumps did. The only question was, where in the park did they belong? That is to say: even though both sides in the ski-jump debate began in the middle ground, they still could not come to an agreement.

Well before Albany created the Adirondack Park in 1892, many different groups had staked their claim to the region. Creating the park, however, invested the Adirondacks with new meaning and made it the legal possession of all New Yorkers. The park designation gained greater significance in the years after World War I, when mass recreation became the rule in the Adirondacks and set up future conflicts between often disenfranchised Adirondackers and often privileged visitors. The 1894 "forever wild" amendment added a new wrinkle, which forced participants in the ensuing contests over land use to grapple with the meaning of wilderness well before the federal Wilderness Act of 1964 had given legitimacy to the notion of a land "untrammeled by man." Although environmental consciousness intensified and became more widespread in the United States after World War II, postwar conflicts in the Adirondacks carried on a story begun during the early twentieth century.

The Olympics crystallized the wilderness debate in the Northeastern United States and shed brighter light on disparate interpretations of modern recreation. Human uses took precedence for all sides in the Olympic debate, but disagreement was rooted in the nature and scale of recreation. How far along the development spectrum should the

Adirondack Park be moved? Should the park be a site for mass recreation or wilderness play? Elevator-aided climbs up ski-jump towers or strenuous exercise? Spectator sports supported by elaborate infrastructure or sightseeing in an open vista? All of the above? Even with the APA and DEC as guides, there were no easy answers to these questions. The only thing on which most participants in the debates could agree was that none of them were happy with the state. Not surprisingly, extinguishing the Olympic flame did not snuff out the conflagrations over land-use practices in the Adirondack Park.

CONCLUSION

"Building Boom in Adirondack Park Tests Laws to Preserve Wilderness." "Will the Adirondacks Secede to Vermont?" "Region and State at Odds over Adirondack Park." "Governor Offers Plan to Preserve Wilderness Site." "Adirondacks in Peril."[1] Headlines such as these appeared in the *New York Times*—but not during the 1970s when the Adirondack Park Agency came into being and almost suffered an early demise. No, the newspaper reported on these developments from 1987 through 1992, at a time when a resurgent land market encouraged new residential construction in the Adirondack Park. During those years, real-estate developers designed and built projects small and clustered enough to escape APA jurisdiction. As a result, the number of single-family dwellings in the Adirondack Park rose from fifty thousand in the late 1960s to seventy thousand twenty years later, enabling the region to welcome some two hundred and fifty thousand seasonal residents a year, about double the region's year-round population.[2] Even though the APA had failed to prevent residential development from reaching new heights, the agency still attracted tremendous animosity. In 1990, for instance, a group of residents formed the Adirondack Solidarity Alliance to study the possibility of seceding from New York State because they claimed to "have no representation" in how the Adirondack Park was managed.[3]

In an attempt to address the pressing problems facing the park, in 1989 Governor Mario Cuomo created the Commission on the Adirondacks in the Twenty-first Century, which went on to formulate a more restrictive land-management plan to control a building boom the APA was powerless to prevent. Ultimately, in 1991 Cuomo introduced watered-down versions of the commission's recommendations that never passed a divided legislature. Reflecting the rightward turn of New York (and American) politics since the late 1960s when the Temporary Study Commission began its work, the centerpiece of Cuomo's aborted legislation was a set of tax incentives to promote preservation. Not even these measures passed the Republican-controlled state senate, which instead proposed a series of bills to abolish the APA.[4] Despite the state's legal victories during the 1970s, many Adirondackers refused to accept the agency's legitimacy, and they built an institutional infrastructure that has continued to stoke anti-APA sentiment into the twenty-first century. As we have seen, the debate over the long-delayed Adirondack Club and Resort in Tupper Lake is clear evidence that intense wilderness politics have continued to roil the region. As a result of such standoffs, the Adirondacks remain trapped in a cycle in which the next big tourist development is touted as the region's savior—that is, until salvation proves illusory and the next recreational project carries the weight of residents' hopes for a better future. It was during the years between 1920 and 1980 that this vicious cycle was established.

During the twentieth century, the state, North Country lawmakers, commercial advocates, and private developers made a devil's bargain with tourism. Even though prodevelopment groups made scapegoats of park administrators for Adirondackers' struggles, the state consistently followed the path of recreational development—even, at times, in violation of the constitution's "forever wild" provision. Thus, the modern wilderness playground in the Adirondacks was not the product of compromise between opposing factions; rather, it was largely the result of the state's default position for most of the twentieth century. Yet even with Albany dedicated to improving the tourism industry in the Adirondacks, the best-case scenario for the many park residents involved sporadic construction and maintenance work and seasonal service jobs. Domesticating the Adirondacks to facilitate outdoor recreation only served to maintain many residents' dependence on visiting vacationers and perpetuate their economic plight. In the process, outsiders gained a greater stake in the park's future, and they tried to direct land-use policy in order to preserve and restore the wild natural environment they enjoyed.

Judging by the continuation of a century-old debate in the Adirondack Park, the state plans enacted from 1920 to 1980 do not provide the clearest roadmap to a more harmonious future. Imposing legibility on the region's land and people turned out to be a quixotic task. Ecological changes, natural energy flows, and unpredictable weather frequently exposed the flaws of state schemes, leaving polluted campgrounds, struggling ski centers, irritated recreation seekers, anxious second-home owners, frustrated businesspeople, and disgruntled Adirondackers in their wake. Continued construction, improved technology, more ambitious recreational programs, and intricate land plans often could not solve old problems, and they also created new ones. The persistent, state-aided pursuit of economic growth through recreational development tended to enrich the few, while it simultaneously fomented more strife and bitterness. Most everyone involved in the Adirondack story has accepted the park as a middle ground combining the features of wilderness and city, and yet environmental politics there have been as combative as anywhere in the United States.

The problem seems to stem less from people's radically different perspectives on the land and more from the fact that few have questioned the centrality of tourism in the Adirondacks and the industry's place in the larger economy. That is, everyone has followed the *same* script. Titled *Battles of the North Country*, Act I begins with development advocates proposing a new recreational project. Next, nature enthusiasts and some landowners challenge the plans, but ultimately the project goes forward in some form somewhere in the Adirondack Park. Nobody is perfectly happy with the resulting middle ground, and most blame the state for failing to protect their interests. Environmentalists lament the loss of another piece of nature, and large numbers of Adirondackers remain impoverished. There is no fade to black, because the story repeats itself again and again. And, as with most sequels, subsequent episodes leave people even more disappointed and perplexed. The common thread connecting this series of conflicts is a tragic flaw that both protagonists and antagonists share. They have failed to question an economic arrangement marked by a fundamental contradiction: rural areas are expected to play catch-up with highly developed cities whose inhabitants demand that parks remain wild spaces for outdoor recreation.

The "here we go again" nature of Adirondack debates was, in large part, the product of the unequal and contentious park-city relationship. Opposing sides in land-use disputes too often began from a position of mutual distrust and antagonism. On one side were many residents resentful of the state and privileged outsiders, both of which wielded significant power in the Adirondack Park. Developers and local politicians, for their part, became experts at exploiting Adirondackers' sense of disenfranchisement to further their own economic and political ends. On the other side were environmental advocates interested in saving nonhuman nature and protecting their own uses of the wilderness. Their reactive activism, which most often emerged to combat new construction projects, offered few constructive alternatives to recreational development, thus reinforcing the notion that environmentalists were meddling obstructionists. By playing that part, they largely accepted the status quo established in the modern wilderness playground and failed to set their sights on the economic realities that forced residents to pin their hopes on the next big development. In its subordinate relationship to the city, the Adirondack Park was trapped in a narrative of progress (as represented by urban areas) and regress (as represented by rural areas) with opposing factions repeating the same arguments over and over.

Regardless of the various players' wishes the narrative is bound to follow a different course in the future, as climate change inexorably reshapes the Adirondacks and redefines human beings' place in the region and on the earth. Will skiing and other winter sports continue to be viable recreational options and sources of revenue?[5] If the winter of 2015–2016 is a sign of things to come, Adirondackers and winter-sports enthusiasts have reason to worry. The 58 inches of snow that fell on Whiteface that winter, far below the average of more than one hundred inches, led to a steep decline in skier days. In addition, cross-country skiers, snowmobilers, and other recreation seekers stayed home instead of filling area hotels, restaurants, and stores. As one local commented, "This winter has seen some crazy weather. I hope to never have another one like it."[6]

Unfortunately, Adirondack winters are not the only part of Adirondack life that is in the throes of significant change as a result of climate change. Rising temperatures threaten to alter the region's waterways and flora and fauna, in the process remaking the park that so many have called home and countless others have called a playground. According to ecologist Jerry Jenkins, the author of *Climate Change in the Adirondacks*, "For the hunters, farmers, hikers and birders, the change in the climate, especially in the past 10 to 15 years, is just too great to write off."[7] What will be left to fight over in future battles of the North Country?

Everyone concerned with the health of the Adirondack Park faces considerable challenges ahead, but therein also lie opportunities. Since environmentalists and most Adirondackers have a shared interest in uncoupling tourism from the competitive growth spiral that enriches the few and spreads development into fragile new environments, they might take aim at a tourism industry in which the poor serve the privileged. For as long as the Adirondack Park resembles a modern wilderness playground, we will see more rancor-filled sequels to the story begun in the 1920s, which will continue to distract from systemic inequality and a warming planet.

Though they reside in a park with a distinctive environment and history, in some ways Adirondackers' experiences have not been unique. They were, and remain, bound up in a larger economic system as volatile and unforgiving as the winter weather atop Whiteface Mountain—the same system that has spawned the climate change that threatens the Adirondack environment. Only with a vision of environmental justice that accounts for the urban-rural divide, widespread economic inequality in tourism and other industries, and the impact of a warming planet can interested parties move forward in a productive and mutually beneficial fashion. Only then will activists on all sides find a road that leads away from the dependency that has shaped the history of the Adirondack State Park.

NOTES

INTRODUCTION: THE ADIRONDACK PARK AS A MODERN WILDERNESS PLAYGROUND

1. Lisa W. Foderaro, "In the Adirondacks, Joy and Fear Over What a Resort May Bring," *New York Times*, February 8, 2012. See also Kim Smith Dedam, "Tupper Developer Eyes Permit Decision Day," *Plattsburgh Press-Republican*, January 15, 2012; David Gibson and Dan Plumley, "A Development Test in Adirondack Park," *New York Times*, January 16, 2012; and Dedam, "Adirondack Decision Delivers Delight," *Plattsburgh Press-Republican*, January 21, 2012.
2. Dedam, "Green Groups Sue over Tupper Resort," *Press Republican*, March 22, 2012; Brian Nearing, "Big Tupper Ski Center Closes," *Albany Times-Union*, September 26, 2012; Adirondack Park Agency, "Appellate Division Affirms APA Approval of ACR Project" (press release), July 3, 2014, http://www.apa.ny.gov/Press/pressrelease.cfm?PressReleaseID=506; Tom Salitsky, "Cheers and Some Sadness as Suit Ends Against Adirondack Club and Resort," *Lake Placid News*, December 28, 2014; Brian Mann, "ACR awaits first sales," *Adirondack Explorer*, May 8, 2015, http://www.adirondackexplorer.org/stories/acr-awaits-first-sales; Adam St. Pierre, Mann, "Hopes for Tupper Lake resort tangled in unpaid bills, back-taxes," North Country Public Radio, October 6, 2016, http://www.northcountrypublicradio.org/news/story/32701/20161006/hopes-for-tupper-lake-resort-tangled-in-unpaid-bills-back-taxes; "Hard road for ACR" (editorial), *Adirondack Daily Enterprise*, January 10, 2017; Ben Gocker, "Breakthrough for Tupper resort as developers buy 5,800 acres," *Adirondack Daily Enterprise*, May 24, 2017; Aaron Cerbone, "APA permits let Tupper resort sell lots," *Adirondack Daily Enterprise*, 7 October 7, 2017.
3. On Adirondack history, see Charles Albert Schleicher, *The Adirondacks: American Playground* (New York: Exposition Press, 1960); Frank Graham Jr., *The*

Adirondack Park: A Political History (New York: Alfred A. Knopf, 1978); Philip G. Terrie, *Forever Wild: Environmental Aesthetics and the Adirondack Forest Preserve* (Philadelphia: Temple University Press, 1985); Alfred Donaldson, *A History of the Adirondacks*, 2 vols. (Fleischmanns, NY: Purple Mountain, 1992); Paul Schneider, *The Adirondacks: A History of America's First Wilderness* (New York: Henry Holt, 1997); Catherine Henshaw Knott, *Living with the Adirondack Forest: Local Perspectives on Land Use Conflicts* (Ithaca: Cornell University Press, 1998); Barbara McMartin, *Perspectives on the Adirondacks: A Thirty-Year Struggle by People Protecting Their Treasure* (Syracuse: Syracuse University Press, 2002); and Terrie, *Contested Terrain: A New History of Nature and People in the Adirondacks*, 2nd ed. (Syracuse: Syracuse University Press, 2008).

4. On hybrid landscapes, see William Cronon, *Nature's Metropolis: Chicago and the Great West* (New York: W. W. Norton, 1991); Cronon, "Introduction: In Search of Nature," in *Uncommon Ground: Rethinking the Human Place in Nature*, ed. Cronon (New York: W. W. Norton, 1996), 25, 52; Cronon, "The Trouble with Wilderness; or, Getting Back to the Wrong Nature," in *Uncommon Ground*, 69–90; Richard White, *The Organic Machine: The Remaking of the Columbia River* (New York: Hill and Wang, 1996); White, "From Wilderness to Hybrid Landscapes: The Cultural Turn in Environmental History," *The Historian* 66, no. 3 (Fall 2004): 557–64.

5. On deindustrialization and the urban crisis, see Karen DeWitt, "Could Upstate NY Cities Go the Way of Detroit?" North Country Public Radio, July 29, 2013, http://www.northcountrypublicradio.org/news/story/22437/20130729/could-upstate-ny-cities-go-the-way-of-detroit; Thomas J. Sugrue, *The Origins of the Urban Crisis: Race and Inequality in Postwar Detroit* Updated edition (Princeton: Princeton University Press, 2014); David Koistinen, *Confronting Decline: The Political Economy of Deindustrialization in Twentieth-Century New England* (Gainesville: University Press of Florida, 2014); Sean Safford, *Why the Garden Club Couldn't Save Youngstown: The Transformation of the Rust Belt* (Cambridge: Harvard University Press, 2009); Guian A. McKee, *The Problem of Jobs: Liberalism, Race, and Deindustrialization in Philadelphia* (Chicago: University of Chicago Press, 2008); Jefferson Cowie and Joseph Heathcott, eds., *Beyond the Ruins: The Meanings of Deindustrialization* (Ithaca: Cornell University Press, 2003); Steven P. Dandaneau, *A Town Abandoned: Flint, Michigan, Confronts Deindustrialization* (Albany: State University of New York Press, 1996).

6. Karl Jacoby, *Crimes against Nature: Squatters, Poachers, Thieves, and the Hidden History of American Conservation* (Berkeley: University of California Press, 2001), especially 2–3, 5–6, and chapters 1–3.

7. Samuel P. Hays, *Conservation and the Gospel of Efficiency: The Progressive Conservation Movement, 1890–1920* (Cambridge: Harvard University Press, 1959). See also Richard Freeman, "The Ecofactory: The United States Forest

Service and the Political Construction of Ecosystem Management," *Environmental History* 7, no. 4 (October 2002: 632–58); Clayton R. Koppes, "Efficiency, Equity, Esthetics: Shifting Themes in American Conservation," in *The Ends of the Earth: Perspectives on Modern Environmental History*, ed. Donald Worster (New York: Cambridge University Press, 1988); Marvin W. Kranz, "Pioneering in Conservation: A History of the Conservation Movement in New York States, 1865–1903" (PhD diss., Syracuse University, 1961); Ellen Frances Stroud, *Nature Next Door: Cities and Trees in the American Northeast* Seattle: University of Washington Press, 2012); Bret Wallach, *At Odds with Progress: Americans and Conservation* (Tucson: University of Arizona Press, 1991).

8. Roderick Frazier Nash, *Wilderness and the American Mind*, 4th ed. (New Haven: Yale University Press, 2001), chapter 10; Robert Righter, *The Battle over Hetch Hetchy: America's Most Controversial Dam and the Making of Modern Environmentalism* (New York: Oxford University Press, 1995).
9. Daniel B. Botkin, *Discordant Harmonies: A New Ecology for the Twenty-First Century* (New York: Oxford University Press, 1990), 9, 6, 7.
10. Paul W. Hirt, *A Conspiracy of Optimism: Management of the National Forests since World War II* (Lincoln: University of Nebraska Press, 1994); Nancy Langston, *Forest Dreams, Forest Nightmares: The Paradox of Old Growth in the Inland West* (Seattle: University of Washington Press, 1996).
11. David Harvey, *The Condition of Postmodernity: An Enquiry into the Origins of Social Change* (Oxford: Basil Blackwell, 1989), 35.
12. James C. Scott, *Seeing Like a State: How Certain Schemes to Improve the Human Condition Have Failed* (New Haven: Yale University Press, 1998), 2 and 3–6.
13. Stephen Skowronek, *Building a New American State: The Expansion of National Administrative Capacities, 1877–1920* (New York: Cambridge University Press, 1982). See also the literature on the Progressive movement, including: Gabriel Kolko, *The Triumph of Conservatism: A Reinterpretation of American History, 1900–1916* (New York: Free Press, 1963); Michael McGerr, *A Fierce Discontent: The Rise and Fall of the Progressive Movement in America, 1870–1920* (New York: Free Press, 2003); Daniel T. Rodgers, *Atlantic Crossings: Social Politics in the Progressive Age* (Belknap of Harvard University Press, 1998); Robert H. Wiebe, *The Search for Order, 1877–1920* (New York: Hill and Wang, 1967).
14. On postwar environmentalism, see Samuel P. Hays with Barbara D. Hays, *Beauty, Health, and Permanence: Environmental Politics in the United States, 1955–1985* (New York: Cambridge University Press, 1987); Samuel P. Hays, *A History of Environmental Politics since 1945* (Pittsburgh: University of Pittsburgh Press, 2000); Robert Gottlieb, *Forcing the Spring: The Transformation of the American Environmental Movement* (Washington: Island Press, 1995); Adam Rome, *The Bulldozer in the Countryside: Suburban Sprawl and the Rise of American Environmentalism* (New York: Cambridge University Press, 2001; Hal Rothman, *The Greening of a Nation? Environmentalism in the United States*

since 1945 (Fort Worth: Harcourt Brace, 1998); Rothman, *Saving the Planet: The American Response to the Environment in the Twentieth Century* (Chicago: Ivan R. Dee, 2000); Philip Shabecoff, *A Fierce Green Fire: The American Environmental Movement*, rev. ed. (Washington: Island Press, 2003); and James Morton Turner, *The Promise of Wilderness: Environmental Politics since 1964* (Seattle: University of Washington Press, 2012). Historians of environmental justice have rightly critiqued the movement for its elitism and disregard for the poor and racial minorities.

15. Turner, "'The Specter of Environmentalism': Wilderness, Environmental Politics, and the Evolution of the New Right," *Journal of American History* 96 (June 2009): 123–148; Turner, *The Promise of Wilderness*. On environmental backlash and conservatism, see also R. McGreggor Cawley, *Federal Land, Western Anger: The Sagebrush Rebellion and Environmental Politics* (Lawrence: University Press of Kansas, 1993); Jacqueline Vaughn Switzer, *Green Backlash: The History and Politics of Environmental Opposition in the U.S.* (Boulder, CO: Lynne Rienner, 1997); and David Helvarg, *The War against the Greens: The "Wise-Use" Movement, the New Right, and Anti-environmental Violence* (San Francisco: Sierra Club Books, 1994).

16. On links between city and countryside, and city and rural park, see Raymond Williams, *The Country and the City* (New York: Oxford University Press, 1973); Cronon, *Nature's Metropolis*; Matthew Gandy, *Concrete and Clay: Reworking Nature in New York City* (Cambridge: MIT Press, 2002); Theodore Catton, *National Park, City Playground: Mount Rainier in the Twentieth Century* (Seattle: University of Washington Press, 2006); David Stradling, *Making Mountains: New York City and the Catskills* (Seattle: University of Washington Press, 2007); Stradling, *The Nature of New York: An Environmental History of the Empire State* (Ithaca: Cornell University Press, 2010); Stroud, *Nature Next Door*; Theodore Steinberg, *Gotham Unbound: The Ecological History of Greater New York* (New York: Simon & Schuster, 2014).

CHAPTER 1: OLYMPIC TRANSFORMATIONS, PART I

1. George Lattimer, compiler, *Official Report: III Olympic Winter Games, Lake Placid, 1932*, (Lake Placid, NY: III Winter Olympic Games Committee, 1932), 123, 125, 167, 241–242, 244, 270.

2. Lattimer, *Official Report*, 37–42; Donald F. Morgan, "Now the White Flash of Winter Sports," *New York Times*, January 10, 1932; Hazel K. Wharton, "Winter Sports Season Begins," *New York Times*, December 22, 1935; George Christian Ortloff and Stephen C. Ortloff, *Lake Placid: The Olympic Years, 1932–1980: A Portrait of America's Premier Winter Resort* (Lake Placid, NY: Macromedia, 1976), chaps. 1–2; and Elizabeth Rider and Nancy Pepin, "Lake Placid: An Overview," chap. 1 in *Lake Placid and the 1980 Winter Games: Community Attitudes*

and the Planning Process, eds. Stephen D. Papson and Alan M. Schwartz (Canton, NY: Environmental Studies Program and North Country Research Center, St. Lawrence University, 1977), 1; David H. Ackerman, *Lake Placid Club, 1895–1980: An Illustrated History* (Lake Placid, NY: The Foundation, 1998), 238–344.

3. Anne LaBastille, "Canoeing through Time: The Eckford Chain," *Adirondack Life* III (Fall 1972): 37–39; Frank Graham Jr., *The Adirondack Park: A Political History* (New York: Alfred A. Knopf, 1978), 28, 31–32; Craig Gilborn, *Adirondack Camps: Homes away from Home, 1850–1950* (Syracuse: Syracuse University Press, 2000), 85–86; Philip Terrie, *Contested Terrain: A New History of Nature and People in the Adirondacks*, 2nd ed. (Syracuse: Syracuse University Press, 2008), 66. On the railroad and middle-class vacationing, see Cindy S. Aron, *Working at Play: A History of Vacations in the United States* (New York: Oxford University Press, 1999), chap. 2.

4. Craig A. Gilborn, *Adirondack Camps: Homes away from Home, 1850–1950* (Syracuse: Syracuse University Press, 2000), 89; Karl Jacoby, *Crimes against Nature: Squatters, Poachers, Thieves, and the Hidden History of Conservation* (Berkeley: University of California Press, 2001), 11–16. For more on the great camps, see Frank Graham Jr., *The Adirondack Park: A Political History* (New York: Alfred A. Knopf, 1978), chap. 6; Jane Eblen Keller, *Adirondack Wilderness: A Story of Man and Nature* (Syracuse: Syracuse University Press, 1980), chap. 13; Philip G. Terrie, *Forever Wild: Environmental Aesthetics and the Adirondack Forest Preserve* (Philadelphia: Temple University Press, 1985), chap. 4; Paul Schneider, *The Adirondacks: A History of America's First Wilderness* (New York: Henry Holt, 1997), chaps. 21–22; and Terrie, *Contested Terrain: A New History of Nature and People in the Adirondacks*, 2nd ed. (Syracuse: Syracuse University Press, 2008), 71–73, 115–123.

5. Terrie, *Forever Wild: Environmental Aesthetics and the Adirondack Forest Preserve* (Philadelphia: Temple University Press, 1985), chap. 6. Samuel Hays defines conservationists as "people who promoted the 'rational' use of resources, with a focus on efficiency, planning for future use, and the application of expertise to broad national problems." Samuel P. Hays, *Conservation and the Gospel of Efficiency: The Progressive Conservation Movement, 1890–1920* (Pittsburgh: University of Pittsburgh Press, 1959), xii.

6. Article VII, section 7, quoted in Paul Schneider, *The Adirondacks: A History of America's First Wilderness* (New York: Henry Holt, 1997), 227. In 1938, the State Constitution was revised and article VII, section 7, became article XIV, section 1. For more on the origins of the Forest Preserve and Adirondack State Park, see Marvin W. Kranz, "Pioneering in Conservation: A History of the Conservation Movement in New York State, 1865–1903" (PhD diss., Syracuse University, 1961); Graham, *The Adirondack Park*, 119–132; Terrie, *Forever Wild*, 68–108; Paul Schneider, *The Adirondacks*, 201–230; Karl Jacoby, 1–78; and Terrie, *Contested Terrain*, 83–105.

7. Aron, *Working at Play*, 160–178; David Strauss, "Toward a Consumer Culture: 'Adirondack Murray' and the Wilderness Vacation," *American Quarterly* 39 (Summer 1987): 270–286; Graham, *The Adirondack Park*, chaps. 4–5; Stacy Ann Keller, *Adirondack Wilderness: A Story of Man and Nature* (Tucson: University of Arizona Press, 1998), chaps. 13–14; Philip G. Terrie, *Forever Wild: Environmental Aesthetics and the Adirondack Forest Preserve* (Philadelphia: Temple University Press, 1985), chaps. 3–4; Charles Brumley, *Guides of the Adirondacks: A History* (Utica: North Country Books, 1994); Paul Schneider, *The Adirondacks: A History of America's First Wilderness* (New York: Henry Holt, 1997), chap. 24; Hallie E. Bond, Joan Jacobs Brumberg, and Leslie Paris, *"A Paradise for Boys and Girls": Children's Camps in the Adirondacks* (Syracuse: Syracuse University Press, 2006); and Terrie, *Contested Terrain*, chaps. 4–6; Dan White, *Under the Stars: How America Fell in Love with Camping* (New York: Henry Holt and Company, 2016), 62–72, 144, 181, 259, 334. On camping in the United States, see Warren J. Belasco, *Americans on the Road: From Autocamp to Motel, 1910–1945* (Baltimore: Johns Hopkins University Press, 1979); Aron, *Working at Play*, chaps. 7–9; Terence Young, "Camping in America: 1869 to the Present," *Arroyo View* 12 (2000): 9; Marguerite Schaffer, *See America First: Tourism and National Identity, 1880–1940* (Washington, D.C.: Smithsonian Books, 2001); James Morton Turner, "From Woodcraft to 'Leave No Trace': Wilderness, Consumerism, and Environmentalism in Twentieth-Century America," *Environmental History* 7 (July 2002): 462–484; Michael B. Smith, "'The Ego Ideal of the Good Camper' and the Nature of Summer Camp," *Environmental History* 11 (January 2006): 70–101; Leslie Paris, *Children's Nature: The Rise of the American Summer Camp* (New York: New York University Press, 2008); Phoebe Kropp, "Wilderness Wives and Dishwashing Husbands: Comfort and the Domestic Arts of Camping in America, 1880–1910," *Journal of Social History* 43 (Fall 2009): 5–30; and Young, "On Camping and Its Equipment," *Environmental History* 15 (January 2010): 120–28; White, *Under the Stars*.
8. Terrie, *Forever Wild*, 67; LaBastille, "Canoeing through Time," 39. For more on the growth of the tourism industry in the Adirondacks, see Charles Albert Schleicher, *The Adirondacks: American Playground* (New York: Exposition, 1960); Graham, *The Adirondack Park*, chap. 6; Jane Eblen Keller, *Adirondack Wilderness: A Story of Man and Nature* (Syracuse: Syracuse University Press, 1980), chap. 11; Karen Ann Dietz, "A Home in the Woods: Summer Life in the Adirondacks," (PhD diss., University of Pennsylvania, 1992); Terrie, *Forever Wild*, chaps. 3–4; Schneider, *The Adirondacks*; chaps. 21–24; Aron, *Working at Play*, 61–62; Jacoby, *Crimes against Nature*, chap. 2; Bryant F. Tolles Jr., *Resort Hotels of the Adirondacks: The Architecture of a Summer Paradise, 1850–1950* (Hanover, NH: University Press of New England, 2003); Terrie, *Contested Terrain*, chap. 6; and Ackerman, *Lake Placid Club*, 9–85.

9. On the modernization of American cities, see David E. Nye, *Electrifying America: Social Meanings of a New Technology, 1880–1940* (Cambridge: MIT Press, 1990); Harold L. Platt, *The Electric City: Energy and the Growth of the Chicago, 1880–1930* (Chicago: University of Chicago Press, 1991); Mark H. Rose, *Cities of Light and Heat: Domesticating Gas and Electricity in Urban America* (University Park, PA: Pennsylvania State University Press, 1995); Nancy Tomes, *Gospel of Germs: Men, Women, and the Microbe in American Life* (Cambridge: Harvard University Press, 1998); Martin V. Melosi, *The Sanitary City: Urban Infrastructure in America from Colonial Times to the Present* (Baltimore: Johns Hopkins University Press, 2000), especially part 2; Peter Baldwin, "How Night Air Became Good Air, 1776–1930," *Environmental History* 8 (July 2003): 412–429. On the back-to-nature movement, see Peter J. Schmitt, *Back to Nature: The Arcadian Myth in Urban America* (New York: Oxford University Press, 1969).
10. Jacoby, *Crimes against Nature*, chaps. 2 and 3; Terrie, *Contested Terrain*, chaps. 2 and 6.
11. Jacoby, *Crimes against Nature*, 39–41.
12. Jacoby, *Crimes against Nature*, 39–45; Terrie, *Contested Terrain*, 123–24.
13. Terrie, *Contested Terrain*, 115–116.
14. Ralph G. Hamlin and Meredith B. Givens, "Shifting Occupational Patterns," chap. 6 in *Recent Social Trends in the United States: Report of the President's Committee on Social Trends* (New York: McGraw-Hill, 1933), 1:268, 269, 282–283, 286–289, 299–303; Robert S. Lynd and Alice C. Hanson, "The People as Consumers," chap. 17 in *Recent Social Trends*, 2:889; J. F. Steiner, "Recreation and Leisure Time Activities," chap. 18 in *Recent Social Trends*, 2: 912–916, 919, 954–955, 957; Cindy S. Aron, *Working at Play: A History of Vacations in the United States* (New York: Oxford University Press, 1999); Lynn Dumenil, *The Modern Temper: American Society and Culture in the 1920s* (New York: Hill & Wang, 1995); Steven J. Ross, *Working-Class Hollywood* (Princeton: Princeton University Press, 1997).
15. Horace M. Albright, *The Birth of the National Park Service: The Founding Years, 1913–1933* (Salt Lake City: Howes Brothers, 1985; Alfred Runte, *National Parks: The American Experience*, 3rd ed. (Lincoln: University of Nebraska Press, 1987); Dwight F. Rettie, *Our National Park System: Caring for America's Greatest Natural and Historic Treasures* (Urbana: University of Illinois Press, 1995); Richard West Sellars, *Preserving Nature in the National Parks: A History* (New Haven: Yale University Press, 1997); Marguerite Schaffer, *See America First: Tourism and National Identity, 1880–1940* (Washington, Smithsonian Institution Press, 2001); David Louter, *Windshield Wilderness: The Automobile and the Meaning of National Parks in Washington State* (Seattle: University of Washington Press, 2006); Tom Lewis, *Divided Highways: Building the Interstate*

Highways, Transforming American Life (New York: Viking, 1997); Owen Geutfrund, *Twentieth-Century Sprawl: Highways and the Reshaping of the American Landscape* (New York: Oxford University Press, 2004).

16. Allen R. Coggins, "The Early History of Tennessee's State Parks: 1919–1956," *Tennessee Historical Quarterly* 43 (1984): 295–315; Rebecca Conrad, *Places of Quiet Beauty: Parks, Preserves, and Environmentalism* (Iowa City: University of Iowa Press, 1997); Thomas R. Cox, *The Park Builders: A History of State Parks in the Pacific Northwest* (Seattle: University of Washington Press, 1988); Joseph H. Engbeck Jr., *State Parks of California from 1864 to the Present* (Portland, OR: Belding and Graphic Arts Center, 1980); Susan L. Flader, ed. *Exploring Missouri's Legacy: State Parks and Historic Sites* (Columbia: University of Missouri Press, 1992); Blake Andrew Harrison, *The View from Vermont: Tourism and the Making of an American Rural Landscape* (Burlington: University of Vermont Press, 2006; Martin Kleinsorge, *Exploring Colorado State Parks*, 2nd ed. (Niwot: University Press of Colorado, 1997); Ney C. Landrum, ed., *Histories of the Southeastern State Park Systems* (Association of Southeastern Park State Park Directors, 1992); Landrum, *The State Park Movement in America: A Critical Review* (Columbia: University of Missouri Press, 2004); Jay M. Price, *Gateways to the Southwest: The Story of Arizona State Parks* (Tucson: University of Arizona Press, 2004); Sharon Morris Toney, "The Texas State Parks System: An Administrative History," (PhD diss., Texas Tech University, 1995); Washington Parks and Recreation Commission, *A History of Washington State Parks: 1913–1988* (Olympia: Washington State Parks and Recreation Commission, 1988).

17. Committee on the State Park Plan, New York State Association, *A State Park Plan for New York, with a Proposal for the New York Bond Issue* (December 1922), 1–4.

18. Committee on the State Park Plan, *A State Park Plan for New York*, 5. For more on the New York State Association and its park plan, see Robert Caro, *The Power Broker: Robert Moses and the Fall of New York* (New York: Vintage, 1975), 131–135, 166–168.

19. Henry R. Francis, ed., *Proceedings of the Second National Conference on State Parks, May 22–25, 1922* (April 1923), 3–4; Gordon Battle to Nathan Strauss Jr., February 16, 1923, in Robert Moses Papers, series 1: Personal and Library Correspondence, box 2, New York Public Library, Stephen A Schwartzman Building, Manuscript and Archives Division, New York, NY; Moses to Clarence MacCallum, February 25, 1924, in Robert Moses Papers, series 1: Personal and Library Correspondence, box 2; Minutes of the Letchworth Park Committee, December 17, 1924, [3]; May 26, 1925, in American Scenic and Historic Preservation Society Records, Committee Records, box 16, vol. 1, New York Public Library, Stephen A. Schwartzman Building, Manuscript and Archives Division, New York, NY; Minutes of the Board of Trustees, June 18, 1925, 16,

in American Scenic and Historic Society Records, minutes, box 3. On urban parks, see Paul S. Boyer, *Urban Masses and the Moral Order: 1820–1920* (Cambridge: Harvard University Press, 1978); Dominick Cavallo, *Muscles and Morals: Organized Playgrounds and Urban Reform, 1890–1920* (Philadelphia: University of Pennsylvania Press, 1981); Galen Cranz, *The Politics of Park Design: A History of Urban Parks in America* (Cambridge: MIT Press, 1982); David Schuyler, *The New Urban Landscape: The Redefinition of City Form in Nineteenth-Century America* (Baltimore: Johns Hopkins University Press, 1986); and Roy Rosenzweig and Elizabeth Blackmar, *The Park and the People: A History of Central Park* (Ithaca: Cornell University Press, 1992).

20. "Huge Bond Issue for State Parks," *Lake Placid News*, July 4, 1924.
21. "Lake Placid Being Considered for the 1932 Winter Olympics," *New York Times*, January 31, 1928; "Raises $50,000 in Bid for Olympics," *Adirondack Record-Elizabethtown Post*, August 9, 1928; "Tennis a Lure to Adirondack Colony," *New York Times*, August 19, 1928.
22. Lattimer, *Official Report*, 43, 45; "Dr. Godfrey Dewey Urges Prompt and Concerted Action to Get 1932 Winter Olympics for Lake Placid," *Lake Placid News*, March 23, 1928; "Placid to Make Strong Bid for Winter Olympics," *Lake Placid News*, March 30, 1928; Ortloff and Ortloff, *Lake Placid: The Olympic Years*, 49–51.
23. U.S. Bureau of the Census, *Twelfth Census of the United States Taken in the Year 1900, Volume VIII: Manufactures, Part II: States and Territories* (Washington, D.C.: Government Printing Office, 1902), table 6, 596; U.S. Bureau of the Census, *Fourteenth Census of the United States Taken in the Year 1920, Volume IX: Manufactures, 1919* (Washington, D.C., Government Printing Office, 1923) table 2, 971; Terrie, *Contested Terrain*, 74, 84–88.
24. "Dewey to Address Chamber Tuesday Evening," *Lake Placid News*, March 30, 1928; "Enthusiasm Marks Chamber Meet as Body Pledges Efforts to Get 1932 Olympic for Lake Placid," April 6, 1928; "State Acts to Get Winter Olympics," *New York Times*, January 15, 1929; Lattimer, *Official Report*, 45–46.
25. "Huge Bobsled Slide Urged for Lake Placid," *New York Times*, January 13, 1929; "State Acts to Get Winter Olympics"; Lattimer, *Official Report*, 157–158; Peter M. Hopsicker, "Legalizing the 1932 Lake Placid Olympic Bob-run: A Test of the Adirondack Wilderness Culture," *Olympika* XVIII (2009): 105.
26. "State May Build Olympic Bob Run on State Land," *Lake Placid News*, March 29, 1929; "State Passes Bill for Bob-sleigh Run," *New York Times*, March 29, 1929; "Placid Bob-sled Run Bill Becomes Law," *Adirondack Record-Elizabethtown Post*, April 18, 1929; Hopsicker, "Legalizing the 1932 Olympic Bob-run," 106.
27. "Urge Lake Placid for 1932 Olympics," *Adirondack Record-Elizabethtown Post*, January 17, 1929; "Lake Placid Chosen by Olympic Body," *New York Times*, April 11, 1929; "Winter Games Go to Lake Placid; California to Protest," *New York*

Times, April 11, 1929; "Lake Placid Awarded 1932 Winter Olympics," *Lake Placid News*, April 12, 1929; Ortloff and Ortloff, *Lake Placid: The Olympic Years*, 51–55.
28. Lattimer, *Official Report*, 49–52; "Olympic Winter Sports Awarded to Lake Placid," *Ticonderoga Sentinel*, April 18, 1929; and "Dr. Dewey Returns, Tells of Olympics," *New York Times*, May 5, 1929; "How Winter Olympics Were Awarded Lake Placid," *Lake Placid News*, May 17, 1929.
29. Lattimer, *Official Report*, 51.
30. "Dr. Dewey Returns, Tells of Olympics," *New York Times*, May 5, 1929; "Godfrey Dewey Feted at Olympic Testimonial," *Lake Placid News*, May 10, 1929; Lattimer, *Official Report*, 53–54.
31. "Bobsled Sites Discussed," *New York Times*, May 21, 1929; "Placid Plans a Record Slide," *Adirondack Record-Elizabethtown Post*, May 30, 1929.
32. Association quoted in Hopsicker, "Legalizing the 1932 Lake Placid Olympic Bob-run," 106–110.
33. "North Elba Citizens Have Keen Interest in Decision Concerning Bob Run on State Land," *Lake Placid News*, December 6, 1929.
34. *The Association for the Protection of the Adirondacks et al., Respondents, v. Alexander McDonald, Conservation Commissioner of the State of New York, et al., Appellants*, 228 app. div. (NY 15 January 1930); "Bars State Land Sled Run," *New York Times*, January 16, 1930; "Bobsled Run Case Will Be Appealed," *New York Times*, January 17, 1930; "Olympic Plans at Lake Placid Given Set-Back," *Adirondack Record-Elizabethtown Post*, January 23, 1930; Hopsicker, "Legalizing the 1932 Lake Placid Olympic Bob-run," 109–110.
35. *The Association for the Protection of the Adirondacks et al., Respondents, v. Alexander McDonald, Conservation Commissioner of the State of New York, et al., Appellants*, Court of Appeals of New York, 253 NY 234 (NY March 18, 1930); "Committee Forced to Decide on Site for Olympic Bob Run on Privately-Owned Land," *Lake Placid News*, March 21, 1930, 1; "Unexpected Snag in State Park Law," *Essex County Republican*, April 18, 1930; "Olympic Commission Prepares for Decision by Purchase of Private Land for Bobsled Run," *Adirondack Record-Elizabethtown Post*, August 14, 1930; "Olympic Games Barred on State Land," *Adirondack Record-Elizabethtown Post*, August 14, 1930; "Court Protects Forest Preserves," *New York Times*, October 26, 1930. See also Roger C. Thompson, "Politics in the Wilderness: New York's Adirondack Forest Preserve," *Forest History* 6 (Winter 1963): 21; Ortloff and Ortloff, *Lake Placid: The Olympics Years*, 57–59; Graham, *The Adirondack Park*, 186; Hopsicker, "Legalizing the 1932 Lake Placid Olympic Bob-run," 111–112; and Terrie, *Contested Terrain*, 146.
36. "Committee Favors Site on South Mountain," *Lake Placid News*, March 28, 1930; "Decision Pending on Bob Sled Run Site," *Essex County Republican*, March 28, 1930; "Officials Study Bob Run sites and Ski Trails, *Lake Placid News*, April 25, 1930; "Site for Olympic Bobsled Site Has Been Selected,"

Adirondack Record-Elizabethtown Post, May 1, 1930; "Bob Run for Third Winter Games of 1932 Will Be Built on South Mountain," *Lake Placid News*, May 2, 1930; "Expert Approves Bob Sled Site," *Ticonderoga Sentinel*, May 15, 1930; "Olympic Bob Run Site to Be Named Mt. Van Hoevenberg," *Lake Placid News*, July 4, 1930; "Contract Awarded for Bob-Sled Run," *New York Times*, August 2, 1930; "Olympic Games Barred on State Land"; "Bob Run Is Scene of Great Activity Now," *Lake Placid News*, June 12, 1931, 1; "Working on New Club House at Olympic Bob Run," *Lake Placid News*, August 7, 1931, 1; "1,256 Visitors Register at Olympic Bob-Run," July 31, 1931; "The Telephone and Olympic Winter Games," *Adirondack Record-Elizabethtown Post*, June 4, 1931; Lattimer, *Official Report*, 86–87, 157–163.

37. "Start Tearing Down Buildings for Stadium," *Lake Placid News*, January 17, 1930; "State Is Asked for $150,000 Fund to Promote Placid Olympic Winter Games," *Lake Placid News*, March 14, 1930; "Supervisors Inspect Site for 1932 Olympics," *Lake Placid News*, March 14, 1930; "Wide Publicity in Connection with Olympics," *Lake Placid News*, March 14, 1930; "Excavation for Olympic Stadium Under Way," *Lake Placid News*, April 11, 1930; "Legislature Passes Bill Creating Sate Olympic Commission for 1932 Games," *Lake Placid News*, April 18, 1930; "Olympic Body Plans Housing, Transportation," *Lake Placid News*, April 25, 1930; "Dirt Flies at Site of Olympic Stadium," *Lake Placid News*, April 25, 1930; "Olympic Games Commission Has Been Appointed," *Adirondack Record-Elizabethtown Post*, July 17, 1930; III Olympic Winter Games Committee, *Official Report*, 60–61, 147–150.
38. "Governor May Veto Winter Olympics Bill," *New York Times*, February 25, 1931; "Threat of Veto Hangs over the $375,000 Appropriation for 1932 Winter Olympics at Lake Placid," *Ticonderoga Sentinel*, February 26, 1931; Roosevelt quoted in "Governor Signs $375,000 Winter Olympics Bill," *Ticonderoga Sentinel*, March 5, 1931; "Dedication of the $220,000 Olympic Arena at Lake Placid Is Scheduled for Jan. 16, *New York Times*, January 9, 1932, 24; "Lake Placid's New Arena Will Open Saturday," *Ticonderog Sentinel*, January 14, 1932; "Lake Placid Revels as Mercury Falls," *New York Times*, January 17, 1932; Lattimer, *Official Report*, 67–68.
39. "Push Plans for Indoor Ice Arena on Main Street Site; Staked-Out Arena Shows Size," *Lake Placid News*, May 1, 1931.
40. "Lake Placid Thaw Hits Olympics," *New York Times*, January 15, 1932, 26; "Winter Sport Program Hit Hard by Thaw," *Plattsburgh Daily Press*," January 15, 1932; "The Olympics Are On," *Ticonderoga Sentinel*, February 4, 1932; Lattimer, *Official Report*, 150–154, 156–157 (italics added). On the arena, see also Ortloff and Ortloff, *Lake Placid: The Olympic Years*, chap. 7.
41. Lattimer, *Official Report*, 67, 115, 125; "500 Visitors to Be Housed in Pullmans for Olympics," *Essex County Republican*, October 9, 1931.
42. Lattimer, *Official Report*, 87, 93, 96–100, 103; "Wide Publicity in Connection

with Olympics," *Lake Placid News*, March 14, 1930, 1; "State Is Asked for Fund of $150,000 to Promote Placid Olympic Winter Games," *Lake Placid News*, March 14, 1930, 1; "Lake Placid Has Been Given Wide Advertising," *Adirondack Record-Elizabethtown Post*, February 18, 1932. Son the rise of advertising, see Roland Marchand, *Advertising the American Dream: Making the Way for Modernity, 1920–1940* (Berkeley: University of California Press, 1985)
43. Advertisements appeared in the *Adirondack Record-Elizabethtown Post*, February 4, 1932.
44. "Olympic Publicity Helps Season Here," *Lake Placid News*, September 18, 1931, 1.
45. Lattimer, *Official Report*, 70, 72, 150–154; Conservation Department, *Twenty-Fifth Annual Report*, 61, 63; Conservation Department, *Twenty-Eighth Annual Report*, 53; Conservation Department, *Thirtieth Annual Report*, 53–54; Conservation Department, *Thirty-First Annual Report*, 45. On the success of winter sports after the Games, see "At Lake Placid," *New York Times*, February 14, 1932, XX9; "Winter Program at Placid to Continue with Numerous Interesting Events Carded," *Plattsburgh Daily Press*, February 19, 1932; "Winter Sports at Lake Placid Wake Up," *Essex County Republican*, December 2, 1932; "Bobsledding at Lake Placid Is 'King of Sports,'" *Ticonderoga Sentinel*, January 19, 1933; Hazel K. Wharton, "Winter Sports Season Begins," *New York Times*, December 22, 1935; and Michael Strauss, "It All Started with Olympics," *New York Times*, January 22, 1961.
46. Donald F. Morgan, "Now the White Flash of Winter Sports," *New York Times*, January 10, 1932; "Winter Olympics Drew $96,000; Lake Placid Deficit $52,468," *New York Times*, February 17, 1932; "Olympic Gate Slightly Less than Expected," *Plattsburgh Daily Press*, February 18, 1932; "Olympic Deficit Totals $52,500," *Ticonderoga Sentinel*, February 18, 1932; "Decides to Seek Games," *New York Times*, September 14, 1939, 34.

CHAPTER 2: CITIES OF TENTS

1. *The Association for the Protection of the Adirondacks et al., Respondents, v. Alexander McDonald, Conservation Commissioner of the State of New York et al., Appellants*, Court of Appeals of New York, 253 NY 234, (March 19, 1930) (italics added). For discussion of the constitutionality of campsite development, see also [Temporary Study Commission on the Future of the Adirondacks], "ASC Study 7: The Constitutionality of Public Campsite Construction on Forest Preserve Land," (ca. 1969), 4–10, in George Davis Papers, box 4, folder 6, Adirondack Museum Library, Blue Mountain Lake, NY; and Roger C. Thompson, "Politics in the Wilderness: New York's Adirondack Forest Preserve," *Forest History* 6 (Winter 1963): 21.
2. On legibility and high modernism, see James C. Scott, *Seeing Like a State: How Certain Schemes to Improve the Human Condition Have Failed* (New Haven:

Yale University Press, 1998), 2, and 3–6; and David Harvey, *The Condition of Post-Modernity: An Enquiry into the Origins of Social Change* (Oxford: Basil Blackwell, 1989), 35.

3. Philip G. Terrie, *Forever Wild: Environmental Aesthetics and the Adirondack Forest Preserve* (Philadelphia: University of Temple Press, 1985), chap. 6; Terrie, *Contested Terrain: A New History of Nature and People in the Adirondacks*, 2nd ed. (Syracuse: Syracuse University Press, 2008), chaps. 6–7.

4. Conservation Commission, *Ninth Annual Report for the Year 1919* (Albany: JB Lyon, 1920), 116–117. Quoted and discussed in Terrie, *Contested Terrain*, 127.

5. On improved services and rising consumer expectations, see William Leach, *Land of Desire: Merchants, Power, and the Rise of a New Consumer Culture* (New York: Pantheon Books, 1993); Shelley Stamp, *Movie-Struck Girls: Women and Motion Picture Culture after the Nickelodeon* (Princeton: Princeton University Press, 2000), 10–40; and Nancy Tomes, "Merchants of Health: Medicine and Consumer Culture in the United States, 1900–1940," *Journal of American History* 88 (Sept. 2001): 519–47. On consumer culture in the interwar years, see Roland Marchand, *Advertising the American Dream: Making the Way for Modernity, 1920–1940* (Berkeley: University of California Press, 1985); Lizabeth Cohen, *Making a New Deal: Industrial Workers in Chicago, 1919–1939* (New York: Cambridge University Press, 1990); and Lawrence B. Glickman, *A Living Wage: American Workers and the Making of Consumer Society* (Ithaca: Cornell University Press, 1997). On vacationing during the 1920s, see Cindy S. Aron, *Working at Play: A History of Vacations in the United States* (New York: Oxford University Press, 1999), chaps. 7–8.

6. Sign appears in the photograph Moon Hill Campsite at Schroon Lake, c. 1938, New York State Archives, Conservation Department, Photographic Prints and Negatives, 14297-87, No. 2304; "Of Local Interest," *Chateaugay Record*, September 2, 1921; "Party of Eight at Notman Camp," *Adirondack Record-Elizabethtown Post*, March 3, 1922; G.S. Franklin & Co. advertisement, *Chateaugay Record*, June 28, 1925.

7. Terrie agrees that state administrators embraced a "utilitarian" approach to managing Adirondack resources. Many New York conservationists were disciples of Gifford Pinchot, and they believed that an unmanaged forest was a wasted forest. Though he provides a good sense of how administrators gradually accepted recreation as a useful function of the Forest Preserve, he stops short of placing recreational development within the purview of conservation. See Terrie, *Forever Wild*, chap. 6, especially 126–135. On recreation as conservation, see also Rebecca Conrad, *Places of Quiet Beauty: Parks, Preserves, and Environmentalism* (Iowa City: University of Iowa Press, 1997).

8. Conservation Commission, *Tenth Annual Report for the Year 1920* (Albany: JB Lyon, 1921), 140–147; "Camping-Canoeing Made Easy to All," *Tupper Lake Herald*, April 27, 1920. See also, Terrie, *Contested Terrain*, 125–127.

9. Conservation Commission, *Fourteenth Annual Report for the Year 1924* (Albany: JB Lyon, 1925), 127.
10. Minutes of the 12th meeting of the State Council of Parks, June 23, 1925, 10, in Calendars and Minutes of Meetings of the State Council of Parks, 1925–1932, box 1, 1925: Meeting Minutes Folder, New York State Archives, Albany, NY.
11. Conservation Commission, *Fourteenth Annual Report*, 127.
12. Lynn Woods, "An Adirondack Auto Biography," in *Adirondack Life* XXVII (May/June 1996): 37–41; Terrie, *Forever Wild*, 124–125, and chap. 7.
13. William G. Howard quoted in "Campers Warned to Watch Fires and Keep Camps in Sanitary Condition," *Lake Placid News*, June 6, 1924.
14. On campers' own efforts to domesticate campsites during the late nineteenth and early twentieth centuries, see Phoebe Kropp, "Wilderness Wives and Dishwashing Husbands: Comfort and the Domestic Arts of Camping in America, 1880–1910," *Journal of Social History* 43 (Fall 2009): 5–30. The shift we see in the 1920s involved the state beginning to bear the burden of domesticating forestlands for campers on a large scale.
15. "Campers Warned to Watch Fires and Keep Camps in Sanitary Condition," *Lake Placid News*, June 6, 1924. See also Matthias Nicoll Jr., State Commissioner of Public Health, quoted in "Health Office Warns of Dangers Menacing Campers in Water Supply," *Lake Placid News*, July 4, 1924.
16. Conservation Commission, *Fifteenth Annual Report* (Albany: JB Lyon, 1926), 18 (italics added).
17. Conservation Commission, *Fifteenth Annual Report* (Albany: JB Lyon, 1926), 159–162.
18. Charles A. Holmquist to William G. Howard, September 19, 1929, 4–6, Hearthstone Point Camp Folder; Holmquist to Howard, September 19, 1929, 6, State Campsite Battle Ground Folder, in State Park and Campsite Wastewater Disposal Files, box 2, New York State Archives, Albany, NY.
19. Holmquist to Howard, September 10, 1929, 7–8, in State Park and Campsite Wastewater Disposal Files, box 2, State Campsite Eagle Point folder.
20. Holmquist to Howard, September 19, 1929, 9–10, State Park and Campsite Wastewater Disposal Files, box 2, Folder Campsite (abandoned) Owens Pond, New York State Archives, Albany, NY.
21. A. F. Allen, the chief of the Bureau of Camp Sanitation, to William D. Mulholland, August 7, 1936, 1; Holmquist to Mulholland, August 11, 1936, 1–2, in State Park and Campsite Wastewater Disposal Files, box 2, Campsite (abandoned) Connery Pond Folder.
22. A. F. Allen to Mulholland, August 12, 1926, 1–2; Memo Mr. Stalbird to Mr. Allen, May 7, 1938, in State Park and Campsite Wastewater Disposal Files, box 2, State Campsite Crown Point Folder.
23. Conservation Department, *Manual for Caretakers* (1938), 5–6, 8, 4, State Park and Campsite Wastewater Disposal Files, box 2.

24. Conservation Commission, *Fourteenth Annual Report*, 129; Conservation Commission, *Fifteenth Annual Report*, 13; Conservation Department, *Manual for Caretakers*, 16.
25. Conservation Commission, *Fifteenth Annual Report*, 162–163.
26. New York State Conservation Department, *Seventeenth Annual Report for the Year 1927* (Albany: JB Lyon, 1928), 144.
27. Conservation Department, *Seventeenth Annual Report*, 159–64; "Season of Summer Camping Opens with a Rush," *Tupper Lake Herald & Adirondack Mt. Press*, July 14, 1927; *Twentieth Annual Report for the Year 1930* (Albany: JB Lyon, 1930), 62, 64; *Nineteenth Annual Report for the Year 1929* (Albany: JB Lyon, 1930), 21–22.
28. Conservation Department, *Eighteenth Annual Report for the Year 1928* (Albany: JB Lyon, 1929), 58, 60, 72, 76.
29. Conservation Department, *Twentieth Annual Report*, 62, 64, 66, 70.
30. Conservation Department, *Nineteenth Annual Report*, 21–22, 42, 60, 62.
31. Conservation Department, *Twenty-First Annual Report for the Year 1931* (New York: Burland Printing, 1932), 22. On the significance of recreation during the Depression, see Jesse Frederick Steiner, "Challenge of the New Leisure," *New York Times*, September 24, 1933; Aron, *Working at Play*, chap. 9; and Susan Currell, *The March of Spare Time: The Problem and Promise of Leisure in the Great Depression* (Philadelphia: University of Pennsylvania Press, 2005).
32. Conservation Department, *Twenty-Second Annual Report for the Year 1932* (Albany: JB Lyon, 1933), 18, 20, 27, 42, 43, 44. See also "Tourist Business Making Big Gains," *New York Times*, August 8, 1935; and "Adirondack Playspots Beckon," *New York Times*, June 28, 1936.
33. Conservation Department, *Twenty-Second Annual Report for the Year 1932*, 44; *Twenty-Seventh Annual Report for the Year 1937* (Albany: JB Lyon, 1938), 56–57.
34. "Forest Amendment on Ballot Tuesday," *New York Times*, November 6, 1932; "The Recreational Amendment: The Opposing Views Set Out," *New York Times*, October 2, 1932. Lake Placid's Republican club and chamber of commerce endorsed the amendment. See "GOP Club and C of C Endorse Amendment," *Lake Placid News*, November 4, 1932.
35. "The Recreational Amendment: The Opposing Views Set Out."
36. "The Recreational Amendment: The Opposing Views Set Out"; "Recreational Amendment Is Much Opposed," *Plattsburgh Republican*, October 19, 1932; "Recreational Amendment Is Declared Bad," *Plattsburgh Republican*, October 24, 1932; "Organizations Fight Forest Amendment," *Plattsburgh Republican*, November 4, 1932; J. Sidney to the editor, October 31, 1932, in *New York Times*, November 8, 1932; "Relief Issue Wins, Forest Plan Loses," *New York Times*, November 9, 1932.
37. Conservation Department, *Twenty-Third Annual Report for the Year 1933* (Albany: JB Lyon, 1934), 17, 27, 33–34, 40; *Twenty-Fifth Annual Report*, 171, 175.

On the CCC in the Adirondacks, see Graham, *The Adirondack Park*, 190; Terrie, *Forever Wild*, 134; and Terrie, *Contested Terrain*, 155. On the CCC, see also Elmo R. Richardson, "The Civilian Conservation Corps and the Origins of the New Mexico State Park System," *Natural Resources Journal* 6 (April 1966): 6–15; John A. Salmond, *The Civilian Conservation Corps, 1933–1942: A New Deal Case Study* (Durham: Duke University Press, 1967); Perry Merrill, *Roosevelt's Forest Army: A History of the Civilian Conservation Corps, 1933–1942* (Montpelier: By the Author, 1981); John C. Paige, *The Civilian Conservation Corps and the National Park Service, 1933–1942: An Administrative History* (Washington: National Park Service, U.S. Department of the Interior, 1985); Phoebe Cutler, *The Public Landscape of the New Deal* (New Haven: Yale University Press, 1985); Jerrell H. Shofner, "Roosevelt's 'Tree Army,' the Civilian Conservation Corps in Florida," *Florida Historical Quarterly* 65 (1987): 433–56; Susan Schrems, "A Lasting New Deal Legacy: The Civilian Conservation Corps, the National Park Service, and the Development of the Oklahoma State Park System," *Chronicles of Oklahoma* 72 (1994–1995): 368–95; Neil Maher, "A New Deal Body Politic: Landscape, Labor, and the Civilian Conservation Corps," *Environmental History* 7 (October 2002): 435–61; Bryant Simon, "'New Men in Body and Soul': The Civilian Conservation Corps and the Transformation of Male Bodies in the Body Politic," in *Seeing Nature through Gender*, ed. Virginia Scharff (Lawrence: University of Kansas Press, 2003); Joseph M. Speakman, *At Work in Penn's Woods: The Civilian Conservation Corps in Pennsylvania* (University Park: Pennsylvania State University Press, 2006); Maher, *Nature's New Deal: The Civilian Conservation Corps and the Roots of the American Environmental Movement* (New York: Oxford University Press, 2007).

38. *National Geographic* quoted in Woods, "An Adirondack Auto Biography," 43 (italics added).
39. Memorandum Re: Fish Creek Automobile Camp Maintained by Conservation Commission, August 29, 1926; Holmquist to Howard, September 29, 1926, 1, in State Park and Campsite Wastewater Disposal Files, box 2, State Campsite Fish Creek Pond Folder; [Harold J.] Rock Preliminary Report, August 9, 1938; Harold J. Rock Report on Gastro-Enteritis Disturbance Fish Creek Pond Public Camp Site, Town of Santa Clara, Franklin County, August 10, 1938, 3; Report Mr. Stalbird, Dr. Quinlivan Gastro-enteritis outbreak Fish Creek Pond, August 19, 1939; Report on Gastro-enteritis at Fish Creek Pond, August 24, 1939; September 13, 1939 Report on Gastroenteritis, State Park and Campsite Wastewater Disposal Files, box 2, State Campsite Fish Creek Pond Folder.
40. Katherine H. Newbold to Commissioner of Public Health Edward S. Godfrey, July 17, 1940, State Park and Campsite Wastewater Disposal Files, box 2, State Campsite Follensby Pond Folder. The Conservation Department acknowledged its own struggles. See Conservation Department, *Twenty-First Annual Report for the Year 1931* (New York: Burland Printing, 1932), 58.

41. Conservation Department, *Twenty-Eighth Annual Report for the Year 1938* (Albany: JB Lyon, 1939), 48–49; *Twenty-Ninth Annual Report for the Year 1939* (Albany: JB Lyon, 1940), 17, 50–51; Conservation Department, *Thirtieth Annual Report for the Year 1940* (New York: Publisher Printers, 1941), 12, 48–49; *Thirty-First Annual Report for the Year 1941* (New York: Publishers Printing), 14, 33, 41.
42. Conservation Department, *Thirty-Third Annual Report for the Year 1943* (1944), 22; *Thirty-Second Annual Report for the Year 1942* (Albany: Williams Press, 1943), 39–41; *Thirty-Second Annual Report*, 47–48.
43. "War Changes Adirondacks," *New York Times*, June 11, 1943.
44. Conservation Department, *Thirty-Third Annual Report*, 37.

CHAPTER 3: A MOUNTAIN TO CLIMB

1. Description is based on Lee Manchester, "A Ride up the Whiteface Veterans Memorial Highway," *Lake Placid News*, June 10, 2005.
2. "Plan Auto Road Up Whiteface; State Memorial to War Dead," *New York Times*, June 26, 1927; "County Legion Branch Favors Whiteface Rd., *Lake Placid News*, July 1, 1927; "Legion Officers Praise Smith Action on Road," *Lake Placid News*, October 21, 1927; "Saranac Rotarians Favor Whiteface Mountain Highway," *Lake Placid News*, September 23, 1927.
3. "Local Organizations Get Started with Whiteface Rd. Campaign," *Lake Placid News*, June 17, 1927.
4. On scenic highway projects of the 1920s and 1930s, see Harley E. Jolley, *The Blue Ridge Parkway* (Knoxville: University of Tennessee Press, 1969); Robert Caro, *The Power Broker: Robert Moses and the Fall of New York* (New York: Vintage, 1975); Phoebe Cutler, *The Public Landscape of the New Deal* (New Haven: Yale University Press, 1985); Anne Mitchell Whisnant, *Super-Scenic Motorway: A Blue Ridge Parkway History* (Chapel Hill: University of North Carolina Press, 2006); and John C. Miles, *Wilderness in National Parks: Playground or Preserve* (Seattle: University of Washington Press, 2009), 56–58.
5. "Long Isl. Official Endorses Memor'l Rd. Up Whiteface," *Lake Placid News*, September 16, 1927; "Not Much of a Bite" (editorial), *Lake Placid News*, January 13, 1928.
6. "Conservation Association for Whiteface Road," *Lake Placid News*, October 14, 1927; Matthew Henry Hoover to the editor, October 24, 1927, in the *New York Times*, October 30, 1927; "Further Facts as to Why Voters Should Support Whiteface Road Project at November Election" (editorial), *Lake Placid News*, October 21, 1927.
7. "Taking the Mountains on High" (editorial), *New York Times*, April 3, 1927. On recreation as self-improvement, see Roderick Frazier Nash, *Wilderness and the American Mind*, 4th ed. (New Haven: Yale University Press, 2001), chap. 9; Bruce C. Daniels, *Puritans at Play: Leisure and Recreation in Early New*

England (New York: St. Martin's Press, 1995); Cindy S. Aron, *Working at Play: A History of Vacations in the United States* (New York: Oxford University Press, 2001); and Leslie Paris, *Children's Nature: The Rise of the American Summer Camp* (New York: New York University Press, 2008).

8. Maxwell Steinhardt to the editor, October 17, 1927, in the *New York Times*, October 19, 1927.
9. "Social Activities of Lake Placid," *New York Times*, June 28, 1925; "Fight Still Looms on Whiteface Road," *New York Times*, November 13, 1927. On the association's origins and composition, see Philip G. Terrie, *Contested Terrain: A New History of Nature and People in the Adirondacks* (Syracuse: Syracuse University Press, 1997), 115–116.
10. Whiteface Mountain Highway Commission, *First Annual Report of the Whiteface Mountain Highway Commission: 1940* (1940), 3; "Wells Tells Conservation Group Benefits of Proposed Whiteface Memorial Road," *Lake Placid News*, December 9, 1927; "Fight Still Looms over Whiteface Road," *New York Times*, November 13, 1927.
11. Greene quoted in "Col. Greene Urges Larger Road Fund," *New York Times*, March 4, 1929. On Adirondackers' general discomfort with tourism, see Terrie, *Contested Terrain*, 131–132.
12. Whiteface Mountain Highway Commission, *First Annual Report*, 3; "Survey for Whiteface Mountain Memorial Highway Completed—Engineers Leave for Albany," *Lake Placid News*, March 29, 1929; "Bonds Sold, Way Clear for Building of Whiteface Memorial Highway, Advertise Bids for Nov. 5," *Lake Placid News*, October 16, 1931; "Christmas Day 1st Anniversary of Work on Whiteface Mt. Memorial Highway—within Two Miles of Top," *Lake Placid News*, December 23, 1932.
13. "Still 900 Feet to Go on Whiteface Memorial Road," *Lake Placid News*, July 13, 1934; "Construction of Whiteface Road Is Far Advanced," *Ticonderoga Sentinel*, August 25, 1932; "Open Highway Today to Whiteface Peak," *New York Times*, July 20, 1935; "Difficulties Delay Opening for Highway," *Plattsburgh Daily Press*, August 20, 1934.
14. "Road Up Whiteface Opened to Traffic," *New York Times*, July 21, 1935.
15. "Road Up Whiteface Opened to Traffic"; "Stage Coach Leads the Way for Modern Vehicles after Whiteface Mt. Road is Officially Opened," *Lake Placid News*, July 26, 1935; "A Goal Accomplished," *Lake Placid News*, July 26, 1935; "Whiteface Mt. Highway Opened Saturday," *Essex County Republican*, July 26, 1935; "Whiteface Highway Is Opened," *Plattsburgh Daily Press*," July 22, 1935; "17,385 Persons Make Whiteface Trip First Month," *Lake Placid News*, August 23, 1935.
16. "61,478 Up Whiteface since 1935 Opening," *Plattsburgh Daily Press*, July 29, 1936; "The Whiteface Memorial" (editorial), *New York Times*, September 14, 1935; Whiteface Mountain Highway Commission, *Annual Report: 1940*, 12. The

pace of visitors set in 1935 remained fairly consistent through 1940. See 12–13 for the ensuing years.
17. "Dedication of Whiteface Mountain Highway," *Plattsburgh Daily Press*, September 13, 1935.
18. "To Build Campsite on Whiteface," *Lake Placid News*, July 31, 1931; "Camp Site on Whiteface to Accommodate 1,500," *Essex County Republican*, August 7, 1931; "CCC Operations at Whiteface Are Announced," *Plattsburgh Daily Press*, May 24, 1934.
19. "Lehman Vetoes $1,000,000 Whiteface Bill; Favors Delay of Year on Improvements," *New York Times*, May 2, 1935; "Col. Greene Revives His Whiteface Plan," *New York Times*, February 2, 1936.
20. "Col. Greene Revives His Whiteface Plan"; Raymond H. Torrey to the editor, June 9, 1936, in the *New York Times*, June 12, 1936.
21. "Vote for Elevator on Whiteface," *Lake Placid News*, October 18, 1935; "Model Shows Tunnel and Shaft to Complete Whiteface Road Project," *Lake Placid News*, February 7, 1936; "Seek Shelter House Atop Whiteface," *Plattsburgh Daily Press*, March 7, 1936; "Question Right of Construction on Whiteface," *Lake Placid News*, July 3, 1936; "Opens Bid for Tunnel Contract on Whiteface Mt.," *Lake Placid News*, April 16, 1937; "Complete Tunnel on Whiteface Mt.," *Lake Placid News*, June 11, 1937; "Whiteface Work Near Completion," *Plattsburgh Daily Press*, July 1, 1937; "Expect Tunnel and Elevator on Whiteface to be Ready by July 1," *Lake Placid News*, May 27, 1938; "Adirondack Retreats Add New Facilities," *New York Times*, June 12, 1938.
22. "Hills Call," *New York Times*, July 14, 1935.
23. Lithgow Osborne, "Along New York's Ski Trails," *New York Times*, February 24, 1935. On the growing popularity of skiing during the interwar years, see Joseph J. Crowe, "Bunnies, Boomers, and Busy Slopes," *The Conservationist* 18 (December 1963–January 1964): 27–28; Hal Rothman, *Devil's Bargains: Tourism in the Twentieth-Century American West* (Lawrence: University Press of Kansas, 2000), chaps. 7 and 8; Annie Gilbert Coleman, *Ski Style: Sport and Culture in the Rockies* (Lawrence: University Press of Kansas, 2004); John Fry, *The Story of Modern Skiing* (Hanover: University Press of New England, 2006); John B. Allen, *The Culture and Sport of Skiing: From Antiquity to World War II* (Amherst: University of Massachusetts Press, 2007); and Roland Huntford, *Two Planks and a Passion: The Dramatic History of Skiing* (London: Continuum, 2008).
24. On the significance of the 1932 Olympics, see George Lattimer, compiler, *Official Report: III Olympic Winter Games, Lake Placid, 1932* (Lake Placid, NY: III Winter Olympic Games Committee, 1932), 37–42; Lithgow Osborne, "Along New York's Ski Trails," *New York Times*, February 24, 1935; "Whiteface Ski Center's Fate to Be Decided in Nov.," *Lake Placid News*, September 19, 1941; Barnett Fowler, "Winter Sports Right Here in New York," *New York State Conservationist*, vol.

3 (December 1948–January 1949): 3; E. L. Frank, "New York to Open Belleayre Center," *New York Times*, December 4, 1949; and Michael Strauss, "It All Started with the Olympics," *New York Times*, January 22, 1961; Allen, *From Ski Sport to Skiing*, 92–96. On the development of skiing in Lake Placid, see Galen Crane, "Ghost Stories: Remembering Lost Adirondack Ski Areas," *Adirondack Life*, vol. XXVII (November/December 1996): 52–58, 64–65; and David H. Ackerman, *Lake Placid Club: An Illustrated History, 1895–1980* (Lake Placid, NY: Lake Placid Education Foundation, 1998), 238–330; Patricia and Robert Foulke, "The Story of Skiing in the North Country; Whiteface Mountain Ski Center," *Lake George Mirror Magazine* February 13, 2011, accessed March 10, 2012, http://www.lakegeorgemirrormagazine.com/tag/ski-centers/.

25. Lattimer, *Official Report*, 146; Osborne, "Along New York's Ski Trails," *New York Times*, February 24, 1935.

26. Conservation Department, *Twenty-Seventh Annual Report for the Year 1937* (Albany: J. B. Lyon, 1938): 56–57; Frank Elkins, "New Trails Abound for Skiers," *New York Times*, December 13, 1936; "Federal, State Bodies Lend Aid for Improving Skiing Facilities," *New York Times*, December 5, 1937; "Clash on Ski Trails," *New York Times*, July 1, 1938; Hal Burton, "Ski Touring," *New York State Conservationist* 2 (February–March 1948): 10.

27. Conservation Department, *Twenty-Ninth Annual Report for the Year 1939* (Albany: J. B. Lyon, 1940), 50; "Court Test Urged Before State Conference on Old Statute which Limits Skiing Trails," *New York Times*, December 6, 1936; "Ski Trails from Peaks," *New York Times*, December 3, 1939. On the Department's ski trail program, see also Burton, "Ski Touring," 10.

28. Gibbs quoted in "Court Test Urged Before State Conference on Old Statute which Limits Skiing Trails," *New York Times*, December 6, 1936. In the opinion of William G. Mulholland, Conservation Department Director of Lands and Forests, "We feel that trails developed especially for down hill racing, slalom hills or open slopes have no place on the preserve, for such facilities are not consistent with maintaining the wilderness character of the forest preserve." Quoted in "Federal, State Bodies Lend Aid for Improving Skiing Facilities," *New York Times*, December 5, 1937.

29. "Start Drive for New Trail on Whiteface—Necessary for Development of Skiing Center," *Lake Placid News*, November 11, 1938; "To Dedicate Whiteface Trail on Monday," *Lake Placid News*, December 23, 1938; "Standard Rating Given Whiteface Trail," *Lake Placid News*, June 30, 1939. Over in North Creek, where Gore Mountain's privately run facilities beckoned skiers, the state signaled its willingness to develop skiing wherever possible. The Conservation Department cooperated with the Gore Mountain Ski Club in the construction of the three-quarter-mile-long Garnet Trail on Gore. See "Lake Placid," *New York Times*, December 15, 1940.

30. J. Vernon Lamb Jr. quoted in Ed Hale, "The Whiteface Chronicles: Building a World-Class Ski Center from the Bottom Up," *Adirondack Life* XXIV (January/February 1993): 49.
31. "Experiments with Machine to Climb Whiteface," *Lake Placid News*, January 7, 1938; "Amendment Four Allow Trails on Whiteface," *Lake Placid News*, February 14, 1941; "Senator Feinberg and Whiteface Ski Trail," *Plattsburgh Daily Press*, April 1, 1940; "Legislature Passes Feinberg Ski Trail Bill," *Lake Placid News* April 5, 1940.
32. "New England Proud of Winter Sports," *New York Times*, February 14, 1932; "Osborne Favors Whiteface Mt. Amendment," *Lake Placid News*, October 10, 1941. This was a common argument in favor of a ski center on Whiteface. See also "Ski Council to Work for Passage of Amendment," *Lake Placid News*, October 10, 1941; "Essex Co. GOP Committee Wants Amendment 4," *Lake Placid News*, October 24, 1941; "Council of Parks Favors Whiteface Amendment," *Lake Placid News*, October 31, 1941; and "Voters to Decide" (editorial), *Lake Placid News*, October 31, 1941.
33. Conservation Department, *Thirtieth Annual Report for the Year 1940* (New York: Publishers Printers, 1941), 52 (italics added).
34. "Ad'k Mountain Club Opposes New Road Building," *Lake Placid News*, February 2, 1940; "State Project Urged by Skiers For Whiteface Mountain Slopes," *New York Times*, January 28, 1940; "See Possible Development of Ski Center on Whiteface Mt.," *Lake Placid News*, February 23, 1940; "Whiteface Area Ski Council Meets," *Lake Placid News*, March 29, 1940; H. W. Hicks to the sports editor, October 13, 1941, in the *New York Times*, October 18, 1941; *Plattsburgh Daily Press*, October 11, 1941.
35. Opponents' statement quoted in Frank Elkins, "Skis on Whiteface?" *New York Times*, October 26, 1941.
36. Ottomar H. Van Norden quoted in "Approves Ski Trail Amendment," *Lake Placid News*, October 31, 1941 (italics added).
37. "Whiteface Ski Center's Fate To Be Decided in Nov.," *Lake Placid News*, September 19, 1941; Elkins, "Skis on Whiteface?"; "Amendment Four Passed—Now Permissable [*sic*] to Build Huge Ski Trail System on Whiteface Mt.," *Lake Placid News*, November 7, 1941; "The Fourth Amendment" (editorial), *Lake Placid News*, November 7, 1941; "Skiing Officials Laud State Site," *New York Times*, November 23, 1941; "Late Skiing Seen in State Center," *New York Times*, 24 November 1941; "The Governor Signs" (editorial), *Lake Placid News*, April 14, 1944; "Recalling 25 Years on Whiteface," *Lake Placid News*, February 3, 1983.
38. "Conservation Department Makes Public Plans for Whiteface Mt. Ski Center," *Lake Placid News*, December 5, 1941. On the early plans for Whiteface Ski Center and endorsements from interest groups, see Hicks to the sports editor, *New York Times*; Elkins, "Skis on Whiteface?"; Elkins, "Skiing Officials Laud

State Site," *New York Times*, November 23, 1941; and Elkins, "Late Skiing Seen in State Center," *New York Times*, November 24, 1941.

39. "Bill Sets Legal Machinery in Motion for State Ski Center on Whiteface Mt.," *Lake Placid News*, February 6, 1942. See also "The Wrong Viewpoint" (editorial), *Lake Placid News*, March 13, 1942; Hicks to the editor, *Lake Placid News*, March 27, 1942; "Get Action in Albany on Ski Runs on W'Face," *Lake Placid News*, March 12, 1943; "Act Will Provide Ski Development on Whiteface," *Lake Placid News*, March 10, 1944; and "Dewey Approves Bill on Race Bias," *New York Times*, April 13, 1944.

40. Thomas Dewey quoted in "Governor Thomas Dewey Signs Enabling Bill for New Ski Center on Whiteface Mt.," *Lake Placid News*, April 14, 1944; "The Governor Signs" (editorial), *Lake Placid News*, April 14, 1944.

CHAPTER 4: A MOUNTAIN FOR ALL SEASONS?

1. Sally McMullen, "State-owned Skiing Areas Hurting Private Slopes," April 14, 1967, clipping in Adirondack Mountain Authority Administrative Subject Files, box 12, folder 3: Adirondack Mt Authority Publicity (part 2), New York State Archives, Albany, New York.
2. Elkins, "Ski Slopes and Trails," *New York Times*, February 27, 1946; "Halt Ski Development due to Bond Default," *Lake Placid News*, August 30, 1946; "Activity on Whiteface Ski Center Has Interest of Animals," *Lake Placid News*, October 24, 1947; "Pushing Work on New Ski Center," *Lake Placid News*, August 16, 1947.
3. "Ask for Bids on Work at Whiteface Skiing Center," *Lake Placid News*, July 30, 1948; "Ski Lodge Funds Provided," *New York Times*, September 3, 1948.
4. "Draper Named to New State Position," *Lake Placid News*, August 27, 1948; "New York's Ski Plans," *New York Times*, October 10, 1948. On the decline of Gore Mountain, see Arnold H. Volmer Associates, *Gore Mountain Ski Center Feasibility Study: A Report to the Adirondack Mountain Authority* (1962), 6, in Adirondack Mountain Authority Administrative Subject Files, box 1, folder 35, New York State Archives, Albany, NY.
5. E. L. Frank, "New York Skiing Jumps," *New York Times*, December 5, 1948; "Whiteface Ski Center Is Ready but Lacks Snow," *Lake Placid News*, October 22, 1948; "Committee Stops Trying to Dedicate Ski Center," *New York Times*, February 17, 1949; Frank Elkins, "Good Skiing Prevails in Far North for Many Week-End Title Meets," *New York Times*, March 11, 1949.
6. E. L. Frank, "New York to Open Belleayre Center," *New York Times*, December 4, 1949; E. L. Frank, "New York's Skiing," *New York Times*, January 15, 1959; "Governor Thomas Dewey to Dedicate New Skiing Center," *Lake Placid News*, December 16, 1949; "Fire Levels Recreation Lodge at Whiteface Mt. Ski

Development Sunday," *Lake Placid News*, May 11, 1951; "New Loge Under Construction at Skiing Center," *Lake Placid News*, November 2, 1951.
7. "C of C Committee Would Like to See Summer Chair Lift on Whiteface," *Lake Placid News*, August 21, 1953.
8. On the opening of Santa's Workshop, see Paul J. C. Friedlander, "Children's Village," *New York Times*, July 3, 1949; Terrie, *Contested Terrain*, 161.
9. Report for Meeting, Adirondack Mt. Authority, October 13, 1967, 2, in Adirondack Mountain Authority Legal and Administrative Files, box 1, folder 35, New York State Archives, Albany, NY.
10. In 1960, the unemployment rate in the Adirondacks was 11.2 percent, more than double the state's rate of 5.2 percent. See Temporary Study Commission on the Future of the Adirondacks, *The Future of the Adirondacks, Technical Reports: Transportation and the Economy* (Blue Mountain Lake, NY: Adirondack Museum), 21–22.
11. Lucille Dee Rubin, "Central Adirondacks Project," *New York Times*, December 13, 1953; Hale, "The Whiteface Chronicles," *Adirondack Life*, 68.
12. Frank Elkins, "Good Skiing Prevails in Far North for Many Week-End Title Meets," *New York Times*, March 11, 1949; Joe Hackett, "Remembering Whiteface's Opening," *Lake Placid News*, November 28, 2003; Rebecca Stefan, "The 'Olympic Mountain' Celebrates 50th Anniversary," *Lake Placid News*, January 18, 2008; Andy Flynn, "Map of Whiteface Ski Center," *Lake Placid News*, May 9, 2008; Hale, "The Whiteface Chronicles," 49–52; Patricia and Robert Foulke, "The Story of Skiing in the North Country; Whiteface Mountain Ski Center," *Lake George Mirror Magazine*, February 13, 2012, accessed March 10, 2012, http://www.lakegeorgemirrormagazine.com/tag/ski-centers/.
13. William Roden quoted in Minutes of the Meeting of the Adirondack Mountain Authority, May 17, 1961, [1], Adirondack Mountain Authority Administrative Subject Files, box 4, folder 10.
14. "Governor Skis W'Face on Friday" and "Placid Has Busy Winter This Year," *Lake Placid News*, March 25, 1955; "Each of 4 Groups Urge Own Mountain," *Lake Placid News*, November 16, 1956; Michael Strauss, "New York Is Ready," *New York Times*, December 2, 1956; Warren Weaver, "Aid for State Ski Trade," *New York Times*, December 9, 1956; M. S., "Improving New York State's Skiing," *New York Times*, January 13, 1957; Foulke and Foulke, "The Story of Skiing in the North Country," *Lake George Mirror Magazine*.
15. "Harriman Urges Haste on Ski Area," *New York Times*, February 26, 1957; Douglas Dales, "Vote for Vacationing," *New York Times*, April 14, 1957; "Governor Signs Bill for Whiteface," *Lake Placid News*, April 19, 1957; "Will Start New Road at W'Face," *Lake Placid News*, May 17, 1957; "Let Bid for Ski Center Access Road," *Lake Placid News*, May 31, 1957; "Will Open Bids on Ski Trails," *Lake Placid News*, June 7, 1957; "Bids Opened for Work on Ski Trails," *Lake Placid*

News, June 21, 1957; "Low Bidder for Trails at W'Face," *Lake Placid News*, October 4, 1957.

16. "Whiteface Debut," *New York Times*, December 8, 1957; Van H. Seagraves, "Whiteface Mountain Ski Center," *The Conservationist* 13 (December 1958–January 1959): 20–22; "To Dedicate Whiteface SC Saturday," *Lake Placid News*, January 24, 1958; Michael Strauss, "Whiteface: The Empire State's New Ski Realm," *New York Times*, January 26, 1958; Hale, "The Whiteface Chronicles," *Adirondack Life*, 71.
17. Roger Tubby, "A Success at Whiteface," *New York Times*, November 30, 1958.
18. James Loeb, "The Adirondacks from a Chairlift," *New York Times*, May 11, 1958; Tubby, "A Success at Whiteface," *New York Times*, November 30, 1958. A 1959 pamphlet written by the Adirondack Mountain Authority compared the chairlift to "a magic carpet opening a scenic wonderland of tumbling brooks, rolling slopes and Adirondack wildlife." Adirondack Mountain Authority, "Whiteface Memorial Highway and Aerial Chairlifts" brochure (1959), in Adirondack Mountain Administrative Subject Files, box 5, folder 11, New York State Archives, Albany, NY.
19. "Ski Reports Explained by Art Draper," *Lake Placid News*, February 13, 1959.
20. "Building New Beginners Area at Whiteface," *Lake Placid News*, July 31, 1959; Walter Carlson, "Whiteface Ski Center a Success," *New York Times*, March 13, 1960; Harold G. Wilm to Ralph E. Dawson, December 12, 1960; and Lloyd A. Maeder to Harold G. Wilm, November 10, 1960, 1, in Adirondack Mountain Authority Administrative Subject Files, box 1, folder 1.
21. Noreen Grady to Harold G. Wilm, January 7, 1960; Arthur G. Draper to Noreen Grady, January 11, 1960, in Adirondack Mountain Authority Administrative Subject Files, box 1, folder 2.
22. Warren Schlickenrieder to Arthur Draper, February 25, 1960, 1, 2, in Department of Environmental Conservation Division of Water State Park and Campsite Waste Disposal Files, box 2, folder State Parks Whiteface Mt. Authority, New York State Archives, Albany, NY.
23. "Ski Center Bill Signed," *New York Times*, April 30, 1960; Harold G. Wilm to T. Norman Hurd, May 5, 1960, in Adirondack Mountain Authority Administrative Subject Files, box 1, folder 2; "Paine Speaks to GOP Women," *Lake Placid News*, June 3, 1960.
24. William M. Roden to Nelson Rockefeller, January 31, 1963, 2, in Adirondack Mountain Authority Legal and Administrative Files, 1957–1968, box 1, folder 6, New York State Archives, Albany, NY; Michael Strauss, "Sno-Cats Make Whiteface Skiable," *New York Times*, February 16, 1961; "Alpine Event Next Week Will Be Held, Needs Snow," *Lake Placid News*, February 23, 1961. For more on the successful 1959–1960 season, see also M. S., "Snow Puts the Adirondacks Back in Business," *New York Times*, January 24, 1960; and Walter Carlson, "Whiteface Ski Center a Success," *New York Times*, March 13, 1960.

25. Roden to Rockefeller, January 31, 1962, 2; *Adirondack Mountain Authority Annual Report, 1960–1961*, Exhibit A, 2, in Adirondack Mountain Authority Legal and Administrative Files, box 1, folder 6.
26. John H. Wikoff letter printed in the *Lake Placid News*, June 22, 1961.
27. "Placid Businessmen Organize 'Whiteface Action Committee,'" *Lake Placid News*, June 29, 1961; "Whiteface Action or Not . . . An Editorial" (editorial), *Lake Placid News*, June 29, 1961; "Whiteface Action Committee Defends Stand on Gore," *Lake Placid News*, February 15, 1962.
28. *Adirondack Mountain Authority Annual Report, 1960–1961*, 1–2, in Adirondack Mountain Authority Legal and Administrative Files, box 1, folder 6; Michael Strauss, "Adirondack Alert," *New York Times*, December 3, 1961. Whiteface was not the first center in the Adirondacks to try snowmaking machines. During the early 1950s, skiing sites in Saranac and Kiamesha Lakes began using them. See Flora White, "More Skiing Space for New York," *New York Times*, December 9, 1951.
29. Michael Strauss, "Getting the Jump on Winter," *New York Times*, December 3, 1961.
30. Memo from William M. Roden to H. G. Wilm, September 21, 1962, 1, in Adirondack Mountain Authority Administrative Subject Files, box 5, folder 1; "Top Positions at Whiteface Are Changed," *Lake Placid News*, April 19, 1962.
31. Minutes of the Meeting of the Adirondack Mountain Authority, December 12, 1963, in Adirondack Mountain Authority, box 2, folder 6, New York State Archives, Albany, NY; Michael Strauss, "Nudging Nature," *New York Times*, January 31, 1965; Adirondack Mt. Authority Press Release, December 13, 1966, in Department of Environmental Conservation Executive Office Commissioner and Deputy Commissioner Correspondence and Subject Files, box 42, folder Adirondack Mt. Authority, New York State Archives.
32. Adirondack Mt. Authority Press Release, December 22, 1966, in Department of Environmental Conservation Executive Office Commissioner and Deputy Commissioner Correspondence and Subject Files, box 42, folder Adirondack Mt. Authority; Report for Meeting, Adirondack Mountain Authority, October 13, 1967, in Adirondack Mountain Authority Legal and Administrative Files, box 1, folder 35; "Whiteface Mountain Ski Center General Information," [c. 1971], in Department of Environmental Conservation Executive Office Commissioner and Deputy Commissioner Correspondence and Subject Files, box 49, folder James Preston Div. of Lands and Forests.
33. Adirondack Mountain Authority, "A Proposal for a Co-ordinated Promotion Program for the Gore Mountain-Warren County Region" [c. 1963], 2, in Adirondack Mountain Authority Legal and Administrative Files, box 1, folder 4; Michael Strauss, "Rain Puts Damper on Opening of $3 Million Gore Ski Center," *New York Times*, January 26, 1964; Strauss, "Gore Mountain Revived as Major Skiing Area," *New York Times*, January 26, 1964, XX16. On Gore

Mountain's facilities and design, see William Robbins, "Mountains Inspire Ski Lodge's Shape," *New York Times*, April 4, 1965; Memo from Dick Brenna to Harold Mason, October 1, 1965, 2; Adirondack Mountain Authority, "Gore Mountain" brochure [n.d.], in Adirondack Mountain Authority Administrative Subject Files, box 7, folder 12; and *Report for Meeting, Adirondack Mountain Authority, October 13, 1967*, 3, in Adirondack Mountain Authority Legal and Administrative Files, box 1, folder 35.

34. "Whiteface Expansion Slated," *Lake Placid News*, July 15, 1965; "Whiteface Marks Anniversary," *Lake Placid News*, July 22, 1965; "Plans for Whiteface and Gore Detailed," *Lake Placid News*, July 22, 1965; Michael Strauss, "Whiteface 'Lift,'" *New York Times*, August 1, 1965; Samuel Masterson, "Adirondacks Ready if Nature Cooperates," *New York Times*, December 5, 1965.

35. "Whiteface Mountain Ski Center General Information" [ca. 1971], in Department of Environmental Conservation Executive Office Commissioner and Deputy Commissioner Correspondence and Subject Files, box 49, folder James Preston Div. of Lands and Forests.

36. Gerald Buyce, "Boots, Britches, and Bumps," *The Conservationist* 18 (December 1963–January 1964): 31. On the facilities and expansion of Belleayre Ski Center, see Vollmer Associates, "Belleayre Mountain Ski Center Expanded Facilities for Year Round Use: A Report to the Conservation Department" (1963), in Adirondack Mountain Authority Administrative Subject Files, box 14, folder 1. As an example of this trend of providing comfortable accommodations continuing in the decades ahead, the 1990s, ads for the state-run Gore Mountain facility promised visitors: "Gore Mt. offers Stress Therapy." See *Adirondack Life* XXIII (January/February 1992): 23.

37. Leonard Sloane, "For the $750-Million Skiing Industry, Profits Are on a Lift," *New York Times*, February 12, 1967.

38. Hal Burton to Vic, October 31, 1967, 1–2, in Adirondack Mountain Authority Administrative Subject Files, box 13, folder 21; [Anonymous, with map courtesy of Adirondack Resorts Press], "Hoffman Mountain Proposed Ski Development" flyer [1967], in Adirondack Mountain Authority, "Hoffman Mountain," [n.d.], in Adirondack Mountain Authority Administrative Subject Files, box 13, folder 21; "Amendment No. 2—Blue Ridge Ski Center" (editorial), *Lake Placid News*, November 2, 1967.

39. "The Blue Ridge Ski Center (On Hoffman Mountain) Depends on You; Vote 'Yes' on Amendment 'Two' on November 7th" [1967], in Department of Environmental Conservation Executive Office Commissioner and Deputy Commissioner Correspondence and Subject Files, box 18, folder Hoffman Mountain, 1967.

40. Conservation Department Forest Preserve Committee, *Annual Report of the Forest Preserve Committee* (September 30, 1964), in Department of Environmental Conservation Executive Office Commissioner and Deputy

Commissioner Correspondence and Subject Files, box 38, folder Forest Preserve; Adirondack Mountain Authority, "Hoffman Mt. Ski Development," in Adirondack Mountain Authority Administrative Subject Files, box 17, folder 3; "Gov. Rockefeller Publicly Endorses Hoffman Mt. Ski Center," *Warrensburg-Lake George News*, October 26, 1967; *Adirondack Life: Supplement to the "Warrensburg-Lake George News,"* March 9, 1967, in Department of Environmental Conservation Executive Office Commissioner and Deputy Commissioner Correspondence and Subject Files, box 18, folder Hoffman Mountain, 1967.

41. Michael Strauss, "Action Ahead in Adirondacks," *New York Times*, December 3, 1967.
42. Minutes of the public hearing of the New York State Temporary Commission to Study the Future of the Adirondacks at Lake George High School, Lake George, NY, August 22, 1969, 173–174, in Administrative Files of the Temporary Study Commission on the Future of the Adirondacks, New York State Archives, Albany, NY. What Fountain considered to be less than enthusiastic support from the state may have stemmed from a comptroller's report that concluded that a new ski center at Hoffman Mountain would steal business from Gore and Whiteface. The Adirondack Mountain Authority, however, disputed the comptroller's findings and, along with the Conservation Department, endorsed the Hoffman amendment. See Adirondack Mountain Authority, "Analyses of the Comptroller's Audit Report of the Adirondack Mountain Authority for the Period November 1, 1962–March 31, 1966" (1966), [1], in Adirondack Mountain Authority Legal and Administrative Files, box 3, folder 14.
43. Adirondack Mountain Club, "The Hoffman Mountain Ski Proposal: Boon or Boondoggle?" (1967), in Adirondack Mountain Authority Administrative Subject Files, box 13, folder 21. On the Adirondack Mountain Club's position, see also "Hoffman Mountain Ski Proposal," *The Southern New York Sportsman: The Magazine of the Southern New York Fish and Game Association*, vol. 18 (November 1967): 12–13, in Department of Environmental Conservation Executive Office Commissioner and Deputy Commissioner Correspondence and Subject Files, box 18, folder Hoffman Mountain, 1967.
44. Michael Strauss, "State Adds Two Ski Centers," *New York Times*, December 1, 1968. In the light of its consistent failure to pay back investment costs, the authority's demise had been discussed for some time; see, for instance, Vic Ostrowidzki, "Stephens Decries Mountain Authority," *Albany Times-Union*, April 1, 1966; and Bill Roden, "Adirondack Sportsman," *Glens Falls Post Star*, October 20, 1966, clippings in Adirondack Mountain Authority Administrative Subject Files, box 12, folder 8.
45. M. S., "Another Peak Open to Motorists," *New York Times*, June 8, 1969. For more on Prospect Mountain Highway, see Hal Burton to Harold G. Wilm, May 24, 1959, in Adirondack Mountain Administrative Subject Files, box 10, folder

3; and Barnett Fowler, "Battle of Prospect Mountain," *Albany Times-Union*, February 21, 1965.

46. "Beauty of Upstate County Hides Poverty of Its Residents," *New York Times*, March 24, 1971. On Essex County's economic struggles, see also Martin Arnold, "Hard Times Clouding Future of Lake Placid and Saranac," *New York Times*, May 30, 1972; and William R. Saunders and Richard E. Prusko in conjunction with Technical Assistance Center, *XIII Olympic Winter Games, Lake Placid, 1980: Economic Impact Study* (August 1977), 3–5, 33–43, and charts 14–20, 24, and table 10.

CHAPTER 5: ADIRONDACK SPRAWL

1. Minutes of the Public hearing of the Temporary Study Commission on the Future of the Adirondacks, Lake George, New York, August 22, 1969, 157–159. On environmentalists' post–World War II concern with the disappearance of open space, see Adam Rome, *The Bulldozer in the Countryside: Suburban Sprawl and the Rise of American Environmentalism* (New York: Cambridge University Press, 2001), esp. chaps. 4–5.
2. Conservation Department, *Thirty-Fourth and Thirty-Fifth Annual Reports*, 13. See also Perry B. Duryea, "From Dreams to Reality," *New York State Conservationist* 1 (August 1946): 2–3, 29.
3. Marion Clawson and Carlton S. Van Doren, eds., *Statistics on Outdoor Recreation, Part I: The Record since 1956* (Washington, D.C.: Resources for the Future, 1958), 38–39, 64; U.S. Department of the Interior National Park Service, *A Study of the Park and Recreation Problem of the United States* (Washington, D.C.: U.S. Government Printing Office, 1941), 112, 212–215; New York State Office of Parks and Recreation, "Keeping Pace: The Postwar Expansion," in *Fifty Years: 1924–1974* (Albany: Natural Heritage Trust, 1975), n.p.; Conservation Department, *Thirty-Sixth Annual Report for the Year 1946* (1947), 32, 33; *Thirty-Ninth Annual Report for the Year 1949* (1950), 38; New York State Conservation Department Division of Parks State Council of Parks, *New York State Parks: Thirtieth Anniversary, 1924–1954* (1954), 26.
4. Temporary Study Commission on the Future of the Adirondacks, "Proposal: Seasonal Residence Study, [ca. 1969], in George Davis Papers, box 4, Adirondack Park Museum Library; Joseph C. Ingraham, "New Route through the Adirondacks," *New York Times*, October 23, 1955. On second-home culture between the two world wars, see Terrie, *Contested Terrain*, 143–145.
5. State of New York Department of Public Works, *The Adirondack Northway: A Study in Depth with Special Emphasis on Glens Falls-Keeseville Segment* (1959), 1, 3, 8, 10, 12, 16, 21–22, 23, 26. See also Public Works, *The Northway and the Adirondack Forest: A Highway Location Study* [1957], 19–23; "Governor Gets 'Info' on Northway," *Lake Placid News*, August 30, 1957; "McMorran Backs the

Northway," *Lake Placid News*, October 30, 1959; and McMorran, "Adirondack Northway," *New York State Conservationist* 2 (October–November 1959): 2–4.

6. State of New York Department of Public Works, *The Adirondack Northway: A Study in Depth with Special Emphasis on Glens Falls-Keeseville Segment* (1959), 1, 3, 8, 10, 12, 21–22, 23, 26. See also Public Works, *The Northway and the Adirondack Forest: A Highway Location Study* [1957], 19–23; "Governor Gets 'Info' on Northway," *Lake Placid News*, August 30, 1957; and "McMorran Backs the Northway," *Lake Placid News*, October 30, 1959.

7. Public Works, *The Adirondack Northway*, 17, 18, 24, 2. On support for route B, see Public Works, *The Northway and the Adirondack Forest*, 10–11; "Essex Garden Club Approves the Northway," *Lake Placid News*, July 31, 1959; "Unions to Work for Ad'k Northway," *Lake Placid News*, July 24, 1959; "Upstate Farmers 'for' Northway," *Lake Placid News*, October 16, 1959; Roger W. Tubby, chairman of the Committee for the Adirondack Northway, to the editor, October 21, 1959, in the *New York Times*, October 27, 1959; Warren Weaver Jr., "Governor Pushes Northway Plea," *New York Times*, October 28, 1959; Committee for the Adirondack Northway, "Your Adirondack Northway: Your Tax Money and Your Future," [1959], in Adirondack Northway Collection, folder MS-66, Adirondack Museum Library; Committee for the Adirondack Northway, "10 Reasons Why You Should Vote YES on the Adirondack Northway Amendment No. 2," [1959], in Adirondack Northway Collection, folder MS-66; William M Roden to Officers, Directors, and Councilmen, [1959], in Northway Committee Records, box 1, Adirondack Museum Library.

8. Arthur G. Draper to the editor, October 4, 1959, in the *New York Times*, October 12, 1959 (italics added). Draper was no doubt encouraged by the superintendent of public works J. Burch McMorran's press release, which read in part: "Skiers and other winter sports enthusiasts will find it easier to reach the slopes and trails of the many winter sports centers in and near New York State's Adirondack Mountains during the coming cold-weather season." See Public Works, "Opening of First Northway Section, Scheduled for October, Will Ease Travel for Skiers, Other Winter Sports Enthusiasts," [1959], 1, in Adirondack Northway Collection, folder MS-66, Adirondack Museum Library, Blue Mountain Lake, NY.

9. On wise use, see John D. Echevarria and Raymond Booth Eby, eds. *Let the People Judge: Wise Use and the Private Property Rights Movement* (Washington, D.C.: Island Press, 1995); and David Helvarg, *The War against the Greens: The "Wise-use" Movement, the New Right, and the Browning of America*, rev. and updated ed. (Boulder, CO: Johnson Books, 2004).

10. "Northway Route Hit," *New York Times*, November 18, 1957; Paul Hillman and Vilbert L. Ostrander, "The Northway Should Go That-a-Way!" *Albany Times-Union*, February 16, 1958; "Northway Route Splits Upstaters," *New York Times*, February 24, 1958; Warren Weaver Jr., "Forest Preserve to Be Issue Again," *New

York Times, October 5, 1958; John B. Oakes, "Conservation: Saving the Wilds," *New York Times*, October 12, 1958; Oakes, "Conservation: Highway Programs," *New York Times*, November 30, 1958; "Offer Stiff Opposition to Ad'k Northway," *Lake Placid News*, July 3, 1959; "Ex-official Scores State Highway Plan," *New York Times*, September 26, 1959; Joseph C. Ingraham, "Northway's First Section Opens this Month," *New York Times*, October 11, 1959; Eleanor R. Crosby, president Federated Garden Clubs of New York State, to the editor, October 12, 1959, in the *New York Times*, October 20, 1959; "'No' on Amendment No. 2" (editorial), *New York Times*, October 23, 1959; "Northway Opposed," *New York Times*, November 2, 1959; Lawrence King, chairman Schenectady Chapter of Adirondack Mountain Club, January 12, 1960, in Northway Committee Records, box 1.

11. Philip W. Ham, "Citizens' Northway Committee press release," January 9, 1959, in Adirondack Northway Collection, folder MS-66; Citizens' Northway Committee, "Vote 'No' on Amendment No. 2," Issued by Eastern New York Chapter of the Nature Conservancy [1959], in Adirondack Northway Collection, folder MS-66.

12. Citizens' Northway Committee, "Vote 'No' on Amendment No. 2"; Ham, "Analysis of Claims in Brochure 'The Northway' issued by DPW, CD, etc." (September 21, 1959), 2, in Northway Committee Records. See also Ham to Wilbur D. Mills, "Statement to Committee Hearing in Washington, D.C. July 22, 1959 on Financing the Highway Construction Fund" (July 17, 1959), in Northway Committee Records, box 1; Citizens' Northway Committee, Press Release, July 22, 1959, 1–2, in Northway Committee Records, box 1.

13. Citizens' Northway Committee, "Which Way Northway? A Choice of No Return," 1959, [2], in Adirondack Northway Collection, folder MS-66; Citizens' Northway Committee, "Bulletin: Strike a Blow for the Freedom of the Wilderness" (July 4, 1959), 1–2, in Northway Committee Records, box 1; Wilderness Society, "Why Spoil the Adirondacks?" *The Living Wilderness*, no. 70 (Autumn 1959), 29, in Northway Committee Records, box 2; Ben Cummings, "Invasion of NY Forest Preserve Is Election Issue," reprinted from the *Argonaut*, published by the Atlantic Chapter of the Sierra Club (May 1959), in Northway Committee Records, box 2. For more on the Wilderness Society's mission to prevent road-building in parks, see Paul Sutter, *Driven Wild: How the Movement against Automobiles Launched the Modern Wilderness Movement* (Seattle: University of Washington Press, 2002).

14. J. Gilbert Maurer and Vilbert L. Ostrander, "The Northway: Lake Champlain Route versus Adirondack Mountain Route," (U.S. Route 9 Improvement Association, Inc., [1959]); Ostrander, "Champlain Northway, the Water-Level Route, vs.: Adirondack Mountain Northway" (1959), in Adirondack Northway Collection, folder MS-66.

15. Charles Grutzner, "Albany-to-Canada Northway Route Wins in City and Upstate," *New York Times*, November 4, 1959; Ham to Mrs. George H. Fonde, December 9, 1959, in Northway Committee Records, box 1; Ham to Nelson Rockefeller, December 17, 1959, in Northway Committee Records, box 2.
16. Conservation Department, *Outdoor Recreation Survey: A Report of the State of New York Conservation Department to Governor Nelson A. Rockefeller* (February 1, 1960), 3, 10, in New York State Department of Environmental Conservation Commissioner's and Subject Files, box 27, New York State Archives, Albany, NY. On the rise of family vacations after World War II, see Susan Sessions Rugh, *Are We There Yet? The Golden Age of American Family Vacations* (Lawrence: University Press of Kansas, 2008).
17. On federal efforts to promote camping in national parks and forests, see John B. Oakes, "Conservation: Highway Programs," *New York Times*, November 30, 1958; John Ise, *Our National Park Policy: A Critical History* (Baltimore: Johns Hopkins University Press, 1961); Joseph Sax, *Mountains without Handrails: Reflections on the National Parks* (Ann Arbor: University of Michigan Press, 1980); Ronald A. Foresta, *America's National Parks and Their Keepers* (Washington: Resources for the Future, 1984); Alfred Runte, *National Parks: The American Experience*, 3rd ed. (Lincoln: University of Nebraska Press, 1987); Richard West Sellars, *Preserving Nature in the National Parks: A History* (New Haven: Yale University Press, 1997); Justin Reich, "Re-Creating the Wilderness: Shaping Narratives and Landscapes in Shenandoah National Park," *Environmental History* 6 (January 2001): 95–117; Mark Daniel Barringer, *Selling Yellowstone: Capitalism and the Construction of Nature* (Lawrence: University Press of Kansas, 2002), chap. 6; Theodore Catton, *National Park, City Playground: Mt. Rainier in the Twentieth Century* (Seattle: University of Washington Press, 2006); David Louter, *Windshield Wilderness: The Automobile and the Meaning of National Parks in Washington State* (Seattle: University of Washington Press, 2006); Ethan Carr: *Mission 66: Modernism and the National Park Dilemma* (Amherst: University of Massachusetts Press, 2007); and John C. Miles, *Wilderness in National Parks: Playground or Preserve* (Seattle: University of Washington Press, 2009), 132–136.
18. Conservation Department, *Outdoor Recreation Survey*, 12, 11 (italics added).
19. On Americans' enjoyment of amenities after World War II, see Samuel P. Hays with Barbara D. Hays, *Beauty, Health, and Permanence: Environmental Politics in the United States, 1955–1985* (New York: Cambridge University Press, 1987).
20. Temporary Study Commission, "ASC Study 7: The Constitutionality of Public Campsite Construction on Forest Preserve Land," [ca. 1969], 3–4, in George Davis Papers, box 4, folder 6.
21. Memo from Harold A. Jerry to Harry W. Albright, November 30, 1970, 2, in Temporary Commission on the Future of the Adirondacks Correspondence

and Meeting Files, 1967–1971, box 1, correspondence: 1970 folder, New York State Archives, Albany, NY; Temporary Study Commission, *The Future of the Adirondack Park* (1970), 27.

22. New York State Conservation Department Division of Lands and Forests, *Report on Proposed Improvements to 18 Selected New York State Campsites* (Glens Falls, NY: Rist-Frost Associates, 1966), 35–42, Adirondack Mountain Authority Administrative Subject Files, box 15, folder 3, New York State Archives, Albany, NY.

23. Memo from Harold J. Dyer to The Adirondack Study Commission Files re: Public Campsites—Major Overloadings, November 18, 1969, in George Davis Papers, box 4, folder 23, Adirondack Museum Library, Blue Mountain Lake, NY. Fish Creek Pond was not alone. In 1964, rangers had turned away almost 90,000 recreation seekers at state campsites. In 1968, for example, Northampton Beach faced an excess of 15,352 and Lake George Battleground 9,586 persons. See Bureau of Forest Recreation, *Annual Report* (1965), [n.p.], in Administrative Files on Establishment and Early Planning of the Adirondack Park Agency, 1969–1990, box 2, New York State Archives, Albany, NY.

24. Adam Rome, *The Bulldozer in the Countryside: Suburban Sprawl and the Rise of American Environmentalism* (New York: Cambridge University Press, 2001); Christopher C. Sellers, *Crabgrass Crucible: Suburban Nature and the Rise of Environmentalism in Twentieth-Century America* (Chapel Hill: University of North Carolina Press, 2012).

25. Conservation Department, *Outdoor Recreation Survey*, 13–14.

26. On zoning, see Jon A. Peterson, *The Birth of City Planning in the United States, 1840–1917* (Baltimore: Johns Hopkins University Press, 2003), 308–317; Kenneth T. Jackson, *Crabgrass Frontier: The Suburbanization of the United States* (New York: Oxford University Press, 1985); Dolores Hayden, *Building Suburbia: Green Fields and Urban Growth, 1820–2000* (New York: Vintage, 2003), 121; Adam Rome, *The Bulldozer in the Countryside: Suburban Sprawl and the Rise of American Environmentalism* (New York: Cambridge University Press, 2001), 130–131; and Christopher C. Sellers, *Crabgrass Crucible: Suburban Nature and the Rise of Environmentalism in Twentieth-Century America* (Chapel Hill: University of North Carolina Press, 2012), chap. 2.

27. Charles Zinser, *The Economic Impact of the Adirondack Private Land Use and Development Plan* (Albany: State University of New York Press, 1980), 20–23.

28. On "wilderness zoning," see John C. Miles, *Wilderness in National Parks: Playground or Preserve* (Seattle: University of Washington Press, 2009), 123–129, and chap. 7. On the idea of wilderness and post–World War II preservation efforts, see Roderick Frazier Nash, *Wilderness and the American Mind*, 4th ed. (New Haven: Yale University Press, 1982); Craig Allin, *The Politics of Wilderness Preservation* (Westport: Greenwood Press, 1982); Michael Cohen, *The*

Pathless Way: John Muir ad the American Wilderness (Madison: University of Wisconsin Press, 1984); Samuel P. Hays with Barbara D. Hays, *Beauty, Health, and Permanence: Environmental Politics in the United States, 1955–1985* (New York: Cambridge University Press, 1987); Max Oelschlaeger, *The Idea of Wilderness: From Prehistory to the Age of Ecology* (New Haven: Yale University Press, 1991); Mark W.T. Harvey, *A Symbol of Wilderness: Echo Park and the American Conservation Movement* (Albuquerque: University of New Mexico Press, 1994); William Cronon, "The Trouble with Wilderness; or, Getting Back to the Wrong Nature," in *Uncommon Ground: Rethinking the Human Place in Nature*, ed. Cronon (New York: W.W. Norton, 1996); Theodore Catton, *Inhabited Wilderness: Indians, Eskimos, and National Parks in Alaska* (Albuquerque: University of New Mexico Press, 1997); Ethan Carr, *Wilderness by Design: Landscape Architecture and the National Park Service* (Lincoln: University of Nebraska Press, 1998); Mark David Spence, *Dispossessing the Wilderness: Indian Removal and the Making of the National Parks* (New York: Oxford University Press, 1999); and Paul Sutter, *Driven Wild: How the Fight against Automobiles Launched the Modern Wilderness Movement* (Seattle: University of Washington Press, 2002).

29. Minutes of the public hearing held by the Temporary Study Commission, Utica, NY, December 11, 1969, 83–84, in Administrative Files of the Temporary Study Commission on the Future of the Adirondacks, New York State Archives, Albany, NY. On the "leave no trace" movement, see James Morton Turner, "From Woodcraft to 'Leave No Trace,'" 462–484.

30. See Conrad L. Wirth, Ben H. Thompson, and Roger Thompson, "A Report on a Proposed Adirondack Mountains National Park" (1967), in New York State Department of Environmental Conservation Commissioner's and Subject Files, box 27.

31. Graham, *The Adirondack Park*, 219–229; Richard A. Liroff and G. Gordon Davis, *Protecting Open Space: Land Use Control in the Adirondack Park* (Cambridge: Ballinger Publishing, 1981), 16–22; Schneider, *The Adirondacks*, 295–296; Terrie, *Forever Wild*, 155; Elizabeth Folwell, "Present at the Creation," *Adirondack Life* XX (July/August 1989): 58; and Terrie, *Contested Terrain*, 165.

32. Leo O'Brien quoted in "Adirondack Study Group Holds Session—Where? In the Adirondacks!" *Essex County Republican*, February 7, 1969. On the work of the Temporary Study Commission, see Graham, *The Adirondack Park*, 236–260; Liroff and Davis, *Protecting Open Space*, 18–22; Terrie, *Forever Wild*, 155–163; Schneider, *The Adirondacks*, 296–297; and Terrie, *Contested Terrain*, 166–173.

33. Temporary Study Commission, *The Future of the Adirondacks, Technical Reports: Transportation and the Economy* (Blue Mountain Lake, NY: Adirondack Museum), 21–22.

34. Minutes of the public hearing held by the Temporary Study Commission on the Future of the Adirondacks at the Town of Webb School Cafeteria, Old Forge, NY, July 8, 1969, 70.
35. Minutes of the public hearing held by the Temporary Study Commission at the Saranac Lake Town Hall, Saranac Lake, NY, February 18, 1970, 21–22.
36. Minutes of the public hearing held by the Temporary Study Commission, Lake George, NY, 125–126.
37. Minutes of the public hearing held by the Temporary Study Commission, Lake George, 41.
38. Minutes of the public hearing held by the Temporary Study Commission at 270 Broadway, New York, NY, January 22, 1970, 58–59. David L. Newhouse, chairman of the Constitutional Council for the Forest Preserve, agreed: "Public campsites, in addition to being provided only on or in the immediate vicinity of existing highways, should offer only the simple or primitive facilities required for considerations for public health and wilderness recreational activities.... Elaborate facilities—electrical and sanitary hookups for trailers, hot showers, lights, or mechanical amusements—should not be provided." See Minutes of the public hearing held by the Temporary Study Commission, Lake George, 56. Nature enthusiast Thomas Kilborn of Rome, New York, was even more blunt: "Like pregnancies, there is no such thing as a little wilderness, it is a wilderness or it is not. Many states have preserves at present, national parks and natural resources and have learned too late that overdevelopment and commercialism have destroyed the environment. Hordes of people have littered, polluted and destroyed the very beauty in the wilderness that they came to see." See Minutes of the public hearing held by the Temporary Study Commission, Utica, 56.
39. Minutes of the public hearing held by the Temporary Study Commission, Lake George, 139–141.
40. According to Gruppe: "years ago when we were still active in wilderness areas, my wife and I were going through an area and we stopped to cook a meal, and in that process we collected enough beer cans and bottles to completely fill the middle section of our canoe, which we carried out." He described another instance: "when we were going along the west branch of the Sacandaga toward the end of the day, we had a new station wagon, and we were practically able to fill the rear section of that station wagon with all kinds of junk left around, and bottles and beer cans." Minutes of the public hearing held by the Temporary Study Commission, Utica, NY, December 11, 1969, 83–84.
41. Minutes of the public hearing held by the Temporary Study Commission, New York, 34–35.
42. Minutes of the public hearing held by the Temporary Study Commission, Lake George, 148–149, 152, 154.
43. Minutes of the public hearing held by the Temporary Study Commission, New York, 6–8.

44. Minutes of the public hearing held by the Temporary Study Commission, Utica, 31–32 (italics added). Robert Lindsay of the Fulton Chain of Lakes Fish and Game Club, agreed with this anti-urban sentiment: "I refuse to prostitute the Adirondacks and make them something odorous and cement. I think they are something forever wild." See Minutes of the public hearing held by the Temporary Study Commission, Old Forge, 18. For his part, Elliot K. Verner, the Adirondack Mountain School's director of the Adirondack Wilderness Camp in Long Lake, lamented the "gross overcrowding" and "artificialization of the wild." See Minutes of the public hearing held by the Temporary Study Commission, Saranac Lake, 49.
45. Temporary Study Commission, *The Future of the Adirondacks* (1970), 14, in Temporary Study Commission Subject, Correspondence, and Meeting Files, box 7, New York State Archives, Albany, NY.
46. Temporary Study Commission, *The Future of the Adirondack Park* (1970), 74–75. On the Commission's internal discussions of skiing in the Adirondacks, see Memo from Harold A. Jerry Jr. to Harry W. Albright Jr., November 30, 1970, 11, box 1, folder Correspondence 1970; and Temporary Study Commission on the Future of the Adirondacks Minutes of Meeting, May 27–28, 1970, in Temporary Study Commission on the Future of the Adirondacks Subject, Correspondence, and Meeting Files, box 3, folder May 27–28 NYC, New York State Archives.
47. Minutes of the public hearing held by the Temporary Study Commission on the Future of the Adirondacks, Old Forge, NY, July 8, 1969, 4, in Adirondack Planning Division Administrative Files of the Temporary Study Commission on the Future of the Adirondacks, New York State Archives, Albany, NY.
48. Minutes of the public hearing held by the Temporary Study Commission on the Future of the Adirondacks, Rochester, NY, December 10, 1969, 14; Chester L. Burton quoted in "Paul Smith's Experts State View on Adirondacks," *Adirondack Daily Enterprise*, February 20, 1970; Orlando B. Potter to the editor, *Adirondack Daily Enterprise*, August 26, 1970.
49. Harold A. Jerry memo to Horace W. Albright, November 30, 1970, 1–2, in Temporary Commission on the Future of the Adirondacks, Subject, Correspondence, and Meeting Files, box 1, folder correspondence: 1970, New York State Archives; John Stock quoted in Richard F. Estes memo to Adirondack Temporary Study Commission Members Re: Feb. 17, 1970 Advisors Meeting, February 24, 1970, 2, in Temporary Study Commission on the Future of the Adirondacks, Subject, Correspondence, and Meeting Files, box 1, folder correspondence: 1970. For John Stock's reflections on the Temporary Study Commission and the creation of the Adirondack Park Agency, see Catherine Henshaw Knott, *Living with the Adirondack Forest: Local Perspectives on Land Use Conflicts* (Ithaca: Cornell University Press, 1998), 171–181.
50. Temporary Study Commission, *The Future of the Adirondack Park* (1970), 9,

26–27. For further discussion of the Temporary Study Commission's recommendations, see Graham, *The Adirondack Park*, 237–241; Charles I. Zinser, *The Economic Impact of the Adirondack Park Private Land Use and Development Plan* (Albany: State University of New York Press, 1980), 8–9; Liroff and Davis, *Protecting Open Space*, 20–22; Schneider, *The Adirondacks*, 196–197; and Terrie, *Contested Terrain*, 167–168.

51. Temporary Study Commission, *The Future of the Adirondack Park*, 26–27.
52. William E. Farrell, "Bill Creating Adirondacks Unit Stalled by Upstate Opposition," *New York Times*, June 2, 1971; "Stafford's Potential Allies" (editorial), *Plattsburgh Press-Republican*, January 7, 1971; Frank Casier quoted in Elizabeth Folwell, "Present at the Creation," *Adirondack Life* XX (July/August 1989): 61.
53. Reflecting on Frank Casier's troubles with the APA, John Stock noted: "[With] Frank Casier . . . [the APA] changed the rules on him in the middle of the game, which cost him a lot of money. Fortunately, he had a lot of money." See Knott, *Living with the Adirondack Forest*, 176.
54. "Planning for Wilderness" (editorial), *Lake Placid News*, January 7, 1971; Roden, "Adirondack Sportsman," *Hamilton County News*, March 11, 1971, clipping in Temporary Study Commission on the Future of the Adirondacks Subject, Correspondence, and Meeting Files, 1967–1971, box 5, folder ASC Files March to 3/31/71; William F. and Janet R. Rochow to Nelson Rockefeller, 16, March 1971, Temporary Study Commission on the Future of the Adirondacks Subject, Correspondence, and Meeting Files, 1967–1971, box 5, folder ASC Files March to 3/31/71; Everett A. Maybell to the editor, *Plattsburgh Press-Republican*, April 4, 1971; Arthur M. Crocker, the president of the Association for the Protection of the Adirondacks, to the editor, April 19, 1971, in the *New York Times*, April 21, 1971; "Ad'k Park" (editorial), *Essex County Republican*, May 21, 1971; "Adirondack Park Agency—Now" (editorial), *New York Times*, May 12, 1971; "Last Chance on the Adirondacks" (editorial), *New York Times*, May 26, 1971; "Time as a Park's Foe" (editorial), *New York Times*, June 3, 1971.
55. Farrell, "Assembly Backs Control of Adirondack Park Land," *New York Times*, June 8, 1971; Farrell, "Bill Creating Adirondacks Unit Stalled by Upstate Opposition," *New York Times*, June 2, 1971; Roy Southworth, "Paine: Control Use of Adirondack Lands," *Plattsburgh Press-Republican*, January 29, 1971; Farrell, "Adirondack State Park Residents Await Development Agency with Mixed Feelings," *New York Times*, August 24, 1971; Folwell, "Present at the Creation," 109, 111–112.
56. Adirondack Park Agency in consultation with the Department of Environmental Conservation, *Adirondack Park State Land Master Plan* (submitted to Governor Nelson A. Rockefeller, June 1, 1972), 6–18. See also "Some Park Areas More Wild than Others; All State Tent Platforms Must Go by 1975," *Adirondack Daily Enterprise*, May 8, 1972; Courtney Jones, "The Agency: A Review," *Adirondack Life* (Fall 1972): 44–47; and Harold Faber, "State Removing

Man-Made Structures, Including Fire Towers, from Adirondack Park Wilderness," *New York Times*, January 2, 1977.
57. Farrell, "Assembly Backs Control of Adirondack Park Land," *New York Times*, June 8, 1971; Farrell, "Bill Creating Adirondacks Unit Stalled by Upstate Opposition," *New York Times*, 2 June 1971; Roy Southworth, "Paine: Control Use of Adirondack Lands," *Plattsburgh Press-Republican*, 29 January 1971; Farrell, "Adirondack State Park Residents Await Development Agency with Mixed Feelings," *New York Times*, August 24, 1971; Folwell, "Present at the Creation," 109, 111–112.

CHAPTER 6: "THERE WAS ONCE AN ADIRONDACK PARK"

1. Citizens to Save the Adirondack Park to the *Sierra Club Bulletin* [1972], in Richard Lawrence Papers, Series 2: APA, 1971–1994, box 13, Adirondack Museum Library, Blue Mountain Lake, New York.
2. "Notes," *Adirondack Life* III (fall 1972): 54; "Adirondack Challenge" (editorial), *New York Times*, March 31, 1972; Norman J. Van Valkenburgh to Harold K. Hochschild, February 5, 1973, in Department of Environmental Conservation Executive Office Commissioner and Deputy Commissioner Correspondence and Subject Files, box 103, folder Adirondack Private Land File, New York State Archives, Albany, NY.
3. Frank Little quoted in David Bird, "Adirondacks Development Stirs Major Ecology Fight," *New York Times*, May 23, 1972.
4. Bird, "Adirondacks Development Stirs Major Ecology Fight."; Evelyn Cropper to the editor, *Plattsburgh Press-Republican*, April 22, 1972; Nelson Bryant, "Wood, Field, and Stream: Conservationists Attack," *New York Times*, May 7, 1972; "Five Groups Opposing Horizon Plan," *Adirondack Daily Enterprise*, August 29, 1972; "Adirondack Development Opposed," *Plattsburgh Press-Republican*, September 1, 1972; Folwell, "Present at the Creation," 112; "5,153 Sign Horizon Petition," *Plattsburgh Press-Republican*, July 3, 1972.
5. Gladys Coleman quoted in Bird, "Adirondacks Development Stirs Major Ecology Fight."
6. Richard Grover quoted in Phil Gallos, "Pro and Anti-Horizon Views Expressed at CSAP Meeting," *Adirondack Daily Enterprise*, September 28, 1972; Bird, "Adirondacks Development Stirs Major Ecology Fight." See also Thomas W. Poster, "Land Development Key Issue Feeding Adirondack Battle," *Plattsburgh Press-Republican*, June 26 1972; "Horizon Project Symbol of Debate over Adirondack Development," *Adirondack Daily Enterprise*, September 7, 1972.
7. Lionel B. Hepburn to the Adirondack Park Agency, [1972], 2, in Richard Lawrence Papers, series 2, box 13. Letters and memorandums received by the APA overwhelmingly opposed Horizon. See St. Lawrence County Environmental Management Council Resolution, March 16, 1972; Allen P. Splete to Richard

Lawrence, March 22, 1972; John P. Seagle to George Davis, April 5, 1972; David Z. Gerhart to Lawrence, April 14, 1972; Zolmon and Ruth Benin to Lawrence, April 14, 1972; Richard G. Myers to Lawrence, April 24, 1972; Brian R. Rivest to Lawrence, April 26, 1972; Peggy A. Overfield to Lawrence, May 15, 1972; Paul E. Dupont to Lawrence, June 1, 1972; John W. Hardy to Lawrence, July 13, 1972. The lone exception was Colton resident and construction worker Robert O'Gorman to Lawrence, April 12, 1972, Richard Lawrence Papers, series 2, box 13.

8. Larry Cole, "Horizon Illegally Cutting Trees, State Says," *Watertown Daily Times*, October 6, 1972, clipping in Richard Lawrence Papers, series 2, box 13; Robert P. Mahoney to the editor, October 13, 1972, *New York Times*, October 23, 1972; "Horizon Draws Blast; Ignores Harris Hearing," *Adirondack Daily Enterprise*, December 8, 1972; Bird, "Adirondacks Development Stirs Major Ecology Fight"; "Horizon Corporation Named in Land Fraud," *Adirondack Daily Enterprise*, March 18, 1975; Grace Lichtenstein, "New Mexico and Arizona Suing Land Concerns over Sales of Lots," *New York Times*, September 19, 1976.

9. "Housing Development Is Outlined at Tupper Lake Chamber Dinner," *Adirondack Daily Enterprise*, September 20, 1971. On the development plans, see also Ton-Da-Lay, Ltd., "Ton-Da-Lay in the Adirondacks," [ca. 1973], in Real-Estate Developments: Ton-Da-Lay, ca. 1973, Vertical Files, Adirondack Museum Library.

10. "Housing Development Is Outlined at Tupper Lake Chamber Dinner"; "Peter Paine says Horizon Issue Will Make or Break Adk. Agency," *Adirondack Daily Enterprise*, June 19, 1972. For Altamont's zoning restrictions, see Fred Tuemmler & Associates, consultants, "Subdivision Regulations for the Town of Altamont, Franklin County, New York, submitted to the Planning Board of the Town of Altamont Dec. 1963," in Ton-Da-Lay Files, Adirondack Park Agency, Ray Brook, New York.

11. Ton-Da-Lay, Ltd., "Ton-Da-Lay in the Adirondacks," 12, 8.

12. Ton-Da-Lay, Ltd. Protective Covenants, [ca. 1972], 1–2, Real-Estate Developments: Ton-Da-Lay. See also Ton-Da-Lay, Ltd., and Palmer & Johnson, "Application to the New York State Department of Environmental Conservation: Ton-Da-Lay, Ltd. Requesting a Public Hearing on a Vacation and Second-Home Development in the Town of Altamont, County of Franklin, State of New York" (July 1972), in Ton-Da-Lay Files.

13. "ECD Has Power over Water System for Ton-De-Lay [sic]," *Adirondack Daily Enterprise*, October 9, 1972; David W. Johnson to Francis V. Carine, December 12, 1972, in Ton-Da-Lay Files; Bird, "Key Decision Near on Adirondack Development," *New York Times*, November 8, 1972. See also "DEC Plans Second-Home Park Probe," *Plattsburgh Press-Republican*, October 9, 1972; "Adirondack Precedent" (editorial), *New York Times*, November 27, 1972; and "Franklin County Backs Ton-Da-Lay," *Adirondack Daily Enterprise*, November 30, 1972.

14. "T. L. Armour Plant Closes," *Adirondack Daily Enterprise*, September 19, 1967; "T.L. Plywood Plant to Close," *Adirondack Daily Enterprise*, August 19, 1969.
15. Lord-Wood/Larson Associates, Inc., *A Resource Management Program: Altamont Properties of Ton-Da-Lay, in the Heart of the Adirondacks: Part II—Economic Impact* (March 1978), 1, 3, in Ton-Da-Lay Files.
16. Bird, "Key Decision Near on Adirondack Development."
17. Gallos, "By Foot," *Lake Placid News*, May 30, 1974.
18. Gallos, "By Foot."
19. Henry L. Diamond, "Statement in Connection with the Application of Ton-Da-Lay, Ltd.," December 5, 1972, 2, in Ton-Da-Lay Files. See also "DEC Explains Hearing Procedure," *Adirondack Daily Enterprise*, December 8, 1972; "Spectators in Hall at Ton-Da-Lay Hearing," *Adirondack Daily Enterprise*, December 6, 1972. During a June 6, 1972, meeting with the DEC, Paparazzo insisted that he would agree to a hearing only on the development's initial phase. See Minutes of the Meeting with Mr. Louis Paparazzo in the New York State Department of Environmental Conservation at Ray Brook, June 6, 1972, submitted by senior stenographer Marjorie L. Bashant, in Ton-Da-Lay Files. For the objection of Paparazzo's attorneys, see Kronish, Lieb, Shainswit, Weiner, and Hellman, "Verified Petition," September 28, 1973, in Ton-Da-Lay Files. On the other hand, the DEC received several requests urging the agency to broaden the hearing to assess the entire development. See Robert C. Townsend to Lawrence, September 27, 1972; William H. Kissel to Terence P. Curran, October 10, 1972; Mary Hall to Lawrence, November 27, 1972; Carlton W. Cooper and Mary S. Cooper to Diamond, November 28, 1972; and Kenneth L. Crowell Statement for Ton-Da-Lay Hearing, December 5, 1972, in Ton-Da-Lay Files.
20. "Ton-Da-Lay Developer Sees His Plans' Nature in Harmony," *Albany Times-Union*, December 7, 1972; "Paparazzo Defends His Vacation Home Plan," *Adirondack Daily Enterprise*, December 7, 1972.
21. "Developer's Plan Sparks Bitter Controversy in Adirondacks," *Adirondack Daily Enterprise*, December 6, 1972; Libby Beyer to the editor, *Plattsburgh Press-Republican*, January 9, 1973.
22. The State of New York commissioned a study highlighting the burdensome pressures imposed on the Long Island environment. According to the New York Citizens' Advisory Committee on Environmental Quality's 1973 report, "Today, Nassau County, to its critics, stands for untrammeled growth. . . . In many areas open space is virtually gone. As traffic jams, multi-laned highways, and smog close in on the suburbanites in Nassau, many are questioning the quality of the growth they have witnessed." See William K. Kelly, ed., *The Use of Land: A Citizens' Guide to Urban Growth: A Task Force Report Funded by the Rockefeller Brothers Fund* (NYL Thomas Y. Crowell Co., 1973), 39, in Department of Environmental Conservation Executive Office Commissioner

and Deputy Commissioner Correspondence and Subject Files, box 108, folder Lands and Forests, New York State Archives. For more context, see Rome, *Bulldozer in the Countryside*.

23. "More Comments on Ton-Da-Lay" (editorial). The newspaper printed Vanderwalker's response, which read in part: "My comparison to the Boston Tea Party was to emphasize my opinion that the people of the Park are as ready to fight for their rights as their ancestors were at Boston. . . . This development's approval or disapproval will have a very serious effect on the people of the North Country." See "Vanderwalker Attacks *Enterprise* Editorial," *Adirondack Daily Enterprise*, December 20, 1972.

24. Evelyn Outcalt, "Adirondack Park Private Land Plan Now Revealed; Restrictions Are Outlined," *Adirondack Daily Enterprise*, December 21, 1972.

25. Richard B. Purdue to the editor, January 15, 1973, *New York Times*, January 26, 1973; "Ecologists Frustrate Sportsmanship, Dolbeck Says," *Plattsburgh Press-Republican*, February 16, 1973; Folwell, "Present at the Creation," 116; M. A. Farber, "Opposition to the Adirondack Plan for Private Land Is Mounting," *New York Times*, January 22, 1973.

26. Farrell, "Adirondack Plan Given by Agency," *New York Times*, March 10, 1973; Folwell, "Present at the Creation," 113, 116–117; "Modified Plan for Adirondacks Going to Legislature This Week," *New York Times*, March 6, 1973; "Adirondacks Can't Wait" (editorial), *New York Times*, March 8, 1973; Farber, "Assembly Delays Adirondack Park Plan," *New York Time*, March 22, 1973; "Adirondack Deadline" (editorial), *New York Times*, March 26, 1973; Alfonso A. Narvaez, "Adirondack Plan Now Faces Delay," *New York Times*, March 28, 1973; "Test for Governor" (editorial), *New York Times*, March 29, 1973; "Governor Vetoes Adirondack Bill," *New York Times*, May 9, 1973.

27. Outcalt, "Compromise Park Plan Readied for Passage," *Adirondack Daily Enterprise*, May 9, 1973; Farber, "Adirondack Compromise Is Reached," *New York Times*, May 9, 1973; "Legislature Passes Private Land Use Plan," *Adirondack Daily Enterprise*, May 15, 1973; Narvarez, "Legislature Passes Bill on Development of the Adirondacks," *New York Times*, May 15, 1973; Farber, "Landmark Fight over Land Use," *New York Times*, May 20, 1973; Farber, "Governor Signs Adirondack Park Bill," *New York Times*, May 23, 1973; "Salvation for the Adirondacks? The What, Where, and How of the APA," *Adirondack Life* IV (fall 1973): 13–15, 46–48; Liroff and Davis, *Protecting Open Space*, 22–39; Folwell, "Present at the Creation," 116–118.

28. Louis Paparazzo, "Statement by Louis Paparazzo, Ton-Da-Lay, Ltd." (January 20, 1973), in Ton-Da-Lay Files; Paparazzo, "DON'T DELEGATE YOUR BASIC RIGHTS," [ca. 1973], in Ton-Da-Lay Files. Paparazzo was not the only one to compare the APA with the Nazi state. Robert R. Purdy, supervisor of the Town of Keene, later wrote that outsiders had decided "that we should be controlled by the Adirondack Gestapo (APA)." See Purdy to the editor, *Plattsburgh*

Press-Republican, May 12, 1977. APA staff member Richard Beamish and impartial observer Peter T. Biesemeyer of State College, Pennsylvania, wrote the *Lake Placid News* to call for participants in the APA debate to use more restraint in their rhetoric, specifically to avoid the Nazi comparisons. See Beamish to the editor, *Lake Placid News*, March 24, 1977; and Biesemeyer to the editor, *Lake Placid News*, March 24, 1977.

29. On opposition to environmentalism, see James Morton Turner, "'The Specter of Environmentalism': Wilderness, Environmental Politics, and the Evolution of the New Right," in *Journal of American History* 96 (June 2009: 123–148. On conservatism's rise, see Lisa McGirr, *Suburban Warriors: The Origins of the New American Right* (Princeton: Princeton University Press, 2001); Sean Wilentz, *The Age of Reagan: A History, 1974–2008* (New York: Free Press, 2008); David R. Farber, *The Rise and Fall of American Conservatism: A Short History* (Princeton: Princeton University Press, 2010); Jill Lepore, *The Whites of Their Eyes: The Tea Party Revolution and the Battle over American History* (Princeton: Princeton University Press, 2010); and Theda Skopcol and Vanessa Williamson, *The Tea Party and the Remaking of Republican Conservatism* (New York: Oxford University Press, 2012).

30. On the Sagebrush Rebellion, see R. McGreggor Cawley, *Federal Land, Western Anger: The Sagebrush Rebellion and Environmental Politics* (Lawrence: University Press of Kansas, 1993).

31. Audrey Casier, wife of Frank Casier, who in March 1976 cofounded a group called the Adirondack Defense League, said of the APA: "They are mercenaries who have brought their alien form of government into the Adirondacks. Our children won't talk to their children. You've got to shun them in any way you can. This is war." See Mary Fiess, "The Defenders," *Adirondack Life* X (March/April 1979): 38.

32. On post-World War II environmentalism and the expanding environmental regime, see Samuel P. Hays, with Barbara D. Hays, *Beauty, Health, and Permanence: Environmental Politics in the United States, 1955–1985* (New York: Cambridge University Press, 1987); Michael Cohen, *The History of the Sierra Club, 1892–1970* (San Francisco: Sierra Club Books, 1988); Robert Gottlieb, *Forcing the Spring: The Transformation of the American Environmental Movement* (Washington, D.C.: Island Press, 1995); Hal Rothman, *The Greening of a Nation? Environmentalism and the United States since 1945* (Fort Worth: Harcourt Brace, 1998); Rome, *Bulldozer in the Countryside*; and Philip Shabecoff, *A Fierce Green Fire: The American Environmental Movement*. Rev. ed. (Washington: Island Press, 2003); and Sellers, *Crabgrass Crucible*.

33. "Hoff Says Vacation-Home Developments Are Burden," *Adirondack Daily Enterprise*, April 6, 1973; "Hoff Says 'Second Homes' Not Advantageous," *Plattsburgh Press-Republican*, April 9, 1973. On residential development in Vermont, see Graham, 239–240, 258; Liroff and Davis, *Protecting Open Space*,

14; and Terrie, *Contested Terrain*, 165. Vail, Colorado, experienced similar consequences as a result of second-home development during the 1960s and 1970s. See Hal K. Rothman, *Devil's Bargains: Tourism in the Twentieth-Century American West* (Lawrence: University Press of Kansas, 1998), chap. 9.

34. Bill Clark, "Heritage Village No Tax Boon," *The Commercial Record*, August 4, 1971, clipping in Ton-Da-Lay Files. A 2011 article in the *New York Times* highlighted the continued graying of Southbury. See Peter Applebome, "Graying Town Builds a Life Beyond Bingo," *New York Times*, December 5, 2011. See also William Yardley, "Otto Paparazzo, Developer of Clustered Housing, Dies at 88," *New York Times*, October 1, 2014.

35. Clark, "Heritage Village No Tax Boon."

36. Transcript of Continued Proceedings at a Public Hearing Held by the New York State Department of Environmental Conservation in the Matter of the Application of Ton-Da-Lay, Ltd., Town of Altamont, Franklin County, for the Construction of a Water Supply and Distribution System to Serve a 1,000-Acre Area in the Northwesterly Corner of the Town of Altamont, etc., Water Supply Application No. 6160 (April 5, 1973), 5,065, 5,080, 5,091, 5,096–5,101, 5,105–5,106, 5,196, in Ton-Da-Lay Files.

37. Francis V. Carine, "Report of Francis V. Carine, Hearing Officer, to Hon. Henry L. Diamond, Commissioner, Department of Environmental Conservation" (July 20, 1973), 10, 12, in Office of Hearings and Mediation Services Records, Department of Environmental Conservation, Albany, New York.

38. George D. Davis to Paparazzo, July 19, 1971, in Ton-Da-Lay Files; Law firm of Kafin and Needleman, "Brief of Intervener-Respondent Sierra Club, in the case of *Ton-Da-Lay, Ltd. and Franklin County, et al., against Henry L. Diamond, as Commissioner of Environmental Conservation, et al., and Sierra Club*," New York State Supreme Court, Appellate Division, Third Department (February 20, 1974), 14–15, 43, 71–72, 74–76, in Ton-Da-Lay Files.

39. Edwin L. Vopelak, DEC Bureau of Water Regulation, to Paparazzo, August 16, 1972, in Ton-Da-Lay Files; Kafin to the DEC Director of Environmental Analysis, August 10, 1972; Kafin to Ronald W. Pederson, first deputy commissioner of the DEC, October 3, 1972; William H. Kissel, APA counsel, to Terence P. Curran, October 10, 1972; DEC, "Possible Economic and Social Factors for Consideration in Ton-Da-Lay Environmental Impact Hearing" (Draft #2), November 17, 1972, in Ton-Da-Lay Files.

40. Diamond, "In the Matter of the Application of Ton-Da-Lay, Ltd. for the Construction of a Water Supply and Distribution System" (August 1, 1973), 10, 12, in Office of Hearings and Mediation Services Records.

41. Diamond, "Application of Ton-Da-Lay," 13, 18–19. See also Department of Environmental Conservation, "Press Release," August 1, 1973, in Ton-Da-Lay Files; Bird, "Environmental Chief Bars Adirondack Home Project," *New York Times*,

August 2, 1973; "Diamond Nixes Ton-Da-Lay," *Plattsburgh Press-Republican*, August 2, 1973; "Ton-Da-Lay Delayed" (editorial), *New York Times*, August 3, 1973; "Saving Room for the Animals," *New York Times*, August 5, 1973; and "DEC Rejects Ton-Da-Lay Development Application," *Lake Placid News*, August 9, 1973.

42. Bird, "Environmental Chief Bars Adirondack Home Project"; "Ton-Da-Lay Files Suit against EPA," *Plattsburgh Press-Republican*, October 12, 1973; "Ton-Da-Lay Appeal Backed," *Plattsburgh Press-Republican*, November 1, 1973; "Board Appropriates $10,000 to Sue DEC," *Adirondack Daily Enterprise*, November 12, 1973; "Ton-Da-Lay Suit," *Adirondack Daily Enterprise*, November 5, 1973; "Ton-Da-Lay Rejection Is Appealed," *Plattsburgh Press-Republican*, March 7, 1974; and Lisa Forrest, "Court Ruling on Ton Da Lay Water Permit Due," *Plattsburgh Press-Republican*, April 22, 1974.

43. Appellate Division of the Supreme Court of the State of New York, In the Matter of *Ton-Da-Lay, Ltd., Petitioner, and Franklin County, et al., Petitioner-Intervener v. Henry L. Diamond, as Commissioner of Environmental Conservation, Respondents, et al., and Sierra Club, Intervener-Respondent* (44 A.D.2d 430 1974), 434, 435, 436, 439, in Ton-Da-Lay Files. See also "Bulletin," *Adirondack Daily Enterprise*, May 16, 1974; "Ton-Da-Lay Rejected, DEC Powers Cut," *Plattsburgh Press-Republican*, May 17, 1974; "Beyond the Ton-Da-Lay Issue" (editorial), *Plattsburgh Press-Republican*, May 21, 1974; and "Ton-Da-Lay Ready to Start," *Adirondack Daily Enterprise*, May 20, 1974. The state appealed the decision, but higher courts dismissed the DEC's appeals, and by May 9, 1975, the case had been resolved. See "New York Will Appeal Ton-Da-Lay Decision," *Adirondack Daily Enterprise*, June 3, 1974; "ENCON Seeks to Reverse Rulings on Ton-Da-Lay," *Adirondack Daily Enterprise*, July 17, 1974; Outcalt, "Ton-Da-Lay Wins Appellate Decision," *Adirondack Daily Enterprise*, November 20, 1974; "Sierra Club Seeking $10,000 to Fight Ton Da Lay Project," *Adirondack Daily Enterprise*, December 17, 1974; and "Ton-Da-Lay Wins in Appeals Court," *Adirondack Daily Enterprise*, May 9, 1975. As a result of the court's rulings in the Ton-Da-Lay case, in summer 1975 the state legislature and Governor Hugh Carey passed a law giving the DEC the broad power to determine a project's "public necessity" based on its environmental impact. See "Carey Seeks More Power," *Adirondack Daily Enterprise*, April 24, 1975; "Ton-Da-Lay Prompts Senate Bill," *Adirondack Daily Enterprise*, July 11, 1975; and "Ton-Da-Lay Suit Inspires New Law," *Adirondack Daily Enterprise*, August 4, 1975.

44. "New York State Sued for $36 Million in Park Zoning Battle," *Adirondack Daily Enterprise*, December 11, 1974; Bird, "Suit Challenges Adirondack Plan," *New York Times*, December 12, 1974; "Horizon Suit 'No Surprise' to State," *Plattsburgh Press-Republican*, December 13, 1974; "First Legal Test for APA" (editorial), *Plattsburgh Press-Republican*, December 17, 1974.

45. New York State Court of Claims, *Horizon Adirondack Corporation, Claimant, against The State of New York, Defendant* (Claim No. 58949, Motion No. M-18464, 1976), 10, 18, 19, in Adirondack Park Agency Records, box 3, Adirondack Museum Library. On the *Horizon* case, see also Liroff and Davis, *Protecting Open Space*, 168–170.
46. "Jordan River Estates" [1975], 1–3, in Adirondack Park Records, box 23, folder Ton-Da-Lay, Adirondack Museum Library.
47. Paparazzo to Richard A. Persico, September 9, 1974, in Ton-Da-Lay Files. Paparazzo was responding to: Persico to Paparazzo, May 23, 1974; and Richard F. Estes to Paparazzo, August 23, 1974, in Ton-Da-Lay Files.
48. Richard J. McCormick, District Sanitary Engineer, to Ton-Da-Lay Associates, April 14, 1975, in Ton-Da-Lay Files; Memorandum attached to Paparazzo to John Stock, May 21, 1975, 1–3, in Adirondack Park Agency Records, box 23, folder Ton-Da-Lay; Ronald Cooper, "Affidavit in the Case of *APA against Ton-Da-Lay*" (August 1, 1975), 3–4, 6–7, in Ton-Da-Lay Files; "Subdivision Approves," *Adirondack Daily Enterprise*, May 7, 1975; Patrick E. Quinn, Altamont Town Supervisor, to the Adirondack Park Agency, May 14, 1975, in Ton-Da-Lay Files.
49. Persico to Paparazzo, September 24, 1974, in Ton-Da-Lay Files; Cooper, "Affidavit in the Case of *APA against Ton-Da-Lay*," 2–8; APA in cooperation with DEC, "New York State Wild, Scenic, and Recreational Rivers System: Field Investigation Summary (proposed addition), prepared pursuant to Section 15-2715 of Title 27 of the State Environmental Conservation Law" [ca. 1973], 1–2, in Ton-Da-Lay Files; Adam R. Palmer to Murray Susswein, July 17, 1975; Susswein, "Affidavit in the matter of *Adirondack Park Agency v. Ton-Da-Lay* (August 1, 1975), 3–5, in Ton-Da-Lay Files. See also Estes to Paparazzo, May 1, 1975, in Ton-Da-Lay Files; John Stock to Gordon Davis and Dave Hannaburgh, May 6, 1975, in Adirondack Park Agency Records, box 24, folder Ton-Da-Lay.
50. Lefkowitz, "Verified Complaint in the matter of *Adirondack Park Agency v. Ton-Da-Lay*" (August 1, 1975), 3–4, 7, in Ton-Da-Lay Files. See also "Injunction Sought against Ton-Da-Lay," *Adirondack Daily Enterprise*, August 6, 1975; Charles W. Stickle, Executive Assistant to the Attorney General, "Press Release," August 5, 1975, in Ton-Da-Lay Files.
51. Paparazzo, "Affidavit in *Adirondack Park Agency against Ton-Da-Lay*" (December 8, 1975), 2–10, in Ton-Da-Lay Files. On Ton-Da-Lay's arguments, see also Liroff and Davis, *Protecting Open Space*, 161–162.
52. "State Task Force Named to Restudy Park Agency," *Adirondack Daily Enterprise*, October 14, 1975; "The Complete Text of the New Interview with Mrs. Sawaya," *Lake Placid News*, November 13, 1975; Iver Peterson, "Curb on Building in Park Is Scored," *New York Times*, February 22, 1976; "H'town Calls for APA Abolishment," *Adirondack Daily Enterprise*, December 24, 1975; "Essex

Supervisors Vote to Urge APA Abolition," *Plattsburgh Press-Republican*, February 3, 1976; "Anti APA League Plans Capitol March," *Plattsburgh Press-Republican*, February 3, 1976; "A View of Mountain Pique" (editorial), *Lake Placid News*, December 4, 1975; Lance Zant to the editor, *Lake Placid News*, December 4, 1975.

53. "Resolution Franklin County Legislature," December 5, 1975; "Memorandum Franklin County Legislature," December 5, 1975," in Ton-Da-Lay Files; Kathryn Bigow, "Franklin County to Intervene in Ton-Da-Lay vs. APA Case," *Adirondack Daily Enterprise*, December 10, 1975; "18 Parties Join Ton Da Lay Suit," *Plattsburgh Press-Republican*, January 31, 1976; "Court Order Sought for Ton-Da-Lay OK," *Plattsburgh Press-Republican*, December 18, 1975; "Ton-Da-Lay Case opens," *Adirondack Daily Enterprise*, May 17, 1976; "APA Legality Test with Ton-Da-Lay Set," *Plattsburgh Press-Republican*, March 19, 1976.

54. Jan Plumadore quoted in James H. Leggett, "Outrage High in APA Area, Counsel Tells Court," *Schenectady Gazette*, May 18, 1976, clipping in Ton-Da-Lay Files; Ruth Ehrich Friedman and Ralph Friedman, "Affidavit in the case *Adirondack Park Agency v. Ton-Da-Lay*" (September 10, 1975), 3, in Ton-Da-Lay Files.

55. Guy A. Graves, Jr., Decision in the case *Adirondack Park Agency, Kildare Club and Ro. Courtney Jones of the Adirondack Council against Ton-Da-Lay Associates, etc.* (December 19, 1976), esp. 4–5, in Ton-Da-Lay Files. See also "Judge Decides Trial Required in Suit vs. APA," *Plattsburgh Press-Republican*, October 13, 1976; and "APA Faces Test," *Lake Placid News*, October 14, 1976; Liroff and Davis, *Protecting Open Space*, 161–164. As Ton-Da-Lay and the Attorney General's office prepared their cases, the company initiated a separate suit against the state for $45 million in damages. See Harold Bishop, "Ton-Da-Lay Sues State for Damages," *Adirondack Daily Enterprise*, August 1, 1977; "$45 Million Suit by Ton-Da-Lay," *Plattsburgh Press-Republican*, August 4, 1977. The Court of Claims would, on May 31, 1979, dismiss Ton-Da-Lay's suit because it hadn't been filed in a timely manner. See New York Court of Claims, Memorandum Opinion in *Ton-Da-Lay against State of New York* (February 16, 1978), in Ton-Da-Lay Files; Supreme Court Appellate Division, Third Department, Decision in the matter of *Ton-Da-Lay, Ltd. v. State of New York* (claim # 61312, May 31, 1979), in Ton-Da-Lay Files; and "Developer Treated Fairly, Appeals Court Rules," *Plattsburgh Press-Republican*, June 1, 1979.

56. Charles Decker, "Arson Attempt Snuffed at APA," *Adirondack Daily Enterprise*, October 20, 1976; "2 Accused of Adirondack Arson," *New York Times*, October 21, 1976; "Essex Grand Jury Local indictments," *Adirondack Daily Enterprise*, November 22, 1976; "Gale Sentenced to 60 Days' Jail," *Adirondack Daily Enterprise*, April 5, 1977. APA members faced other threats and acts of violence. Richard Estes was punched in the face while the Agency was embroiled in a

dispute over a Fine resident's property. Another staff member had his tires slashed. In November 1975, a pile of horse manure was dumped outside the office entrance. See Fiess, "The Defenders," 37, 42.

57. Audrey Casier to the editor, *Lake Placid News*, November 17, 1976. For more on the Adirondack Defenders and their mission, see Fleiss, "The Defenders," and Anthony N. D'Elia, *The Adirondack Rebellion: A Political, Economic, and Social Expose of the Adirondack State Park* (Loon Lake, NY: Onchiota Books, 1979).

58. Dave Landolfe to the editor, *Lake Placid News*, December 2, 1976. Soon after Landolfe's letter was published, the Local Government Review Board voted 11–1 for the APA's abolition. The lone dissenting member, Richard B. Purdue, who was quoted earlier in this chapter as a staunch APA opponent who argued that the Agency would make local government "a slave to Albany," resigned from the board, which he claimed had "completely given in to the bullying methods of a vocal pressure group." See "Adirondack Showdown" (editorial), *New York Times*, December 10, 1976.

59. Jeremiah M. Hayes, "Respondent's Brief in *Adirondack Park Agency v. Ton-Da-Lay*" (October 4, 1977), 2, 3–6, in Ton-Da-Lay Files.

60. Palmer, "Answer to Interveners' Complaint in *Adirondack Park Agency v. Ton-Da-Lay*" [1976], 6, in Ton-Da-Lay Files; Hayes, "Respondent's Brief," 15–16, 14. Ton-Da-Lay's attorneys made another charge that was settled in a separate case brought by Wambat Realty. Wambat sought APA approval to build Valmont Village in Black Brook, Clinton County. On November 20, 1974, the APA rejected as incomplete the company's application to subdivide 851 units on 2,224 acres, to be part of a larger development with 10,000 lots on 21,000 acres. The thrust of Wambat's argument (one Ton-Da-Lay would revive) was that the APA violated the rights of Adirondack towns as spelled out in Article IX of the State Constitution and the 1964 Statute of Local Governments. These laws stipulated that the legislature must approve legislation overriding a local law (such as a zoning ordinance) at two consecutive sessions. Since the APA Act passed through only one session, so the argument went, the Agency must be dissolved. On December 31, 1975, the State Supreme Court rejected Wambat's plea. Judge James Gibson concluded that the Adirondack Park was a matter of "substantial state interest" and so land-use restrictions there did not violate home-rule legislation. Gibson also called attention to the necessity of the APA at a time when "local governments in the Adirondack park find it increasingly difficult to cope with the unrelenting pressures for development." See "Wambat Realty Challenges APA in Court Case," *Lake Placid News*, January 3, 1975; "Jordan River Estates," Exhibit C, [1975], 1–2; "Brief of the Defendants on the Constitutional Law Questions," *APA against Ton-Da-Lay* (1975), 1–2, in Ton-Da-Lay Files; New York State Supreme Court of Essex County, Memorandum Decision in *Wambat Realty Corp. against State of New York, Richard F. Estes as*

Assistant Director and Director of Operations of Adirondack Park Agency, Board of Members of Adirondack Park Agency, and Adirondack Park Agency (December 31, 1975), 5–7, 9, in Adirondack Park Agency Records, box 23, Adirondack Museum Library. The court reaffirmed the Wambat decision in spring 1977. See "Agency Courts Invulnerability," *Lake Placid News*, June 2, 1977. Loon Lake Estates, Inc., which sought to develop thirty-five hundred acres in the town of Franklin, Franklin County, also challenged the DEC and APA, claiming that their demands for changes to the company's plans exceeded those agencies' police powers by "taking" the developer's lands without compensation. On September 13, 1975, the State Supreme Court dismissed the charges. See New York State Supreme Court of Essex County, Decision in the matter of *Loon Lake Estates, Inc. v. Adirondack Park Agency, et al.* (September 13, 1975), in *83 Miscellaneous Reports, 2d Series*, 686–694, in Adirondack Park Agency Records, box 3. On the *Wambat* and *Loon Lake Estates* cases, see also Liroff and Davis, *Protecting Open Space*, 107–112 and 160–164.

61. Lefkowitz, "Appellant's Brief in *Adirondack Park Agency v. Ton-Da-Lay*" (October 28, 1977), 5–7, 8–9, 12–13; Lefkowitz, "Appellant's Brief in *Adirondack Park Agency v. Ton-Da-Lay* (August 10, 1977), 18–19, 31–32, 37, 50–53, in Ton-Da-Lay Files. See also Liroff and Davis, *Protecting Open Space*, 162–164.

62. New York State Supreme Court, third department, decision in *Adirondack Park Agency v. Ton-Da-Lay* (February 10, 1978), 3–4, in Ton-Da-Lay Files. See also "APA Constitutional, State Court Rules," *Plattsburgh Press-Republican*, February 10, 1978; and Adirondack Park Agency, "Press Release," February 14, 1978.

63. Lord-Wood/Larson Associates, *A Resource Management Program: Part II—Economic Impact*, 4–6; *A Resource Management Program: Part III—Revenue Projections*, 15. See also Paparazzo to the Adirondack Park Agency, July 31, 1978; and Paparazzo to Robert C. Glennon, January 15, 1979, in Ton-Da-Lay Files.

64. Harold Bishop, "Huge Land Sale," *Adirondack Daily Enterprise*, November 12, 1980; Bishop, "Two Large Land Sales Recorded in Franklin," *Adirondack Daily Enterprise*, January 13, 1981. During the mid-1980s, Cobham sold 1,485 acres to the Swiss lumber firm Zepa Consulting Co. and 3,111 acres to the Nature Conservancy for preservation. See "Swiss Firm Buys Portions of Former Ton-Da-Lay Lands," *Adirondack Daily Enterprise*, April 27, 1984; Shawn Tooley, "Naure Conservancy Purchasing 3,111-Acre Tract in Altamont," *Adirondack Daily Enterprise*, February 12, 1985; Tooley, "Adirondack Conservancy Seeks Control of 75,000-Acre Preserve," *Adirondack Daily Enterprise*, May 3, 1985; Tooley, "Altamont Gets Windfall of $23,000," *Adirondack Daily Enterprise*, August 4, 1986.

65. Both counties lagged behind the state average of $141.05. New York State Division of Budget, *New York State Statistical Yearbook* (Albany, 1970), 77. St. Lawrence County's population rose from 111,239 to 111,991. See Bureau of the

Census, *1970 Census of Population: Vol. 1: Characteristics of Population, Part 1: U.S. Summary* (Washington, D.C.: Department of Commerce, 1973), 1–103.

CHAPTER 7: OLYMPIC TRANSFORMATIONS, PART II

1. Edited transcript of a public hearing upon a project permit for the construction of ninety- and seventy-meter ski jump facilities proposed by the Lake Placid Olympic Organizing Committee at Intervale (Town of North Elba) Essex County, New York, to be used for the 1980 Olympic Winter Games before the Adirondack Park Agency of the State of New York, Lake Placid, New York, November 19, 1976, 43–44 (italics added), in Adirondack Park Agency Records, series 3: MS 80-2, box 17: Olympic Hearings—Proceedings for 70 and 90 meter ski jump, Adirondack Museum Library, Blue Mountain Lake, New York.
2. J. Bernard Fell, interviewed by Benson E. Miller for the XIII Winter Games 1980 Oral History, November 7, 1979, 1–3, in MSS Oral History, folder 2, St. Lawrence University Owen D. Young Library, Frank and Anne Piskor Special Collections Area, Canton, NY; Jean Melious, Jo Ellen Richards, and John Mumford, "Planning for the Olympics," chap. 2 in *Lake Placid and the 1980 Olympic Games: Community Attitudes and the Planning Process*, eds. Stephen D. Papson and Alan M. Schwarts (Canton, NY: Environmental Studies Program and the North Country Research Center, St. Lawrence University, 1977), [1–2]; XIII Winter Olympic Games Committee, *Final Report: XIII Olympic Winter Games, Lake Placid, 1980* (1981), 15–16.
3. "Area Facilities Ideal for Winter Olympics," *Lake Placid News*, September 7, 1967; "Residents Extend Momentous Effort for '76 Olympics," *Lake Placid News*, November 22, 1967; Anthony Ripley, "Colorado Is Cool to Winter Games," *New York Times*, January 23, 1972; "Colorado Drops Winter Games Bid," *New York Times*, November 9, 1972; Bernard Kirsch, "Innsbruck, '64 Site, Gets '76 Olympics," *New York Times*, February 5, 1973; Melious et al., "Planning for the Olympics," [2]; George Carroll, *Winter Olympics: Lake Placid, 1980* (Lake Placid, NY: Barry & Wilson, 1979), 18–30; Alan Schwartz, "The Olympics and the Environment," *Adirondack Life* XI (January/February 1980): 23, 26; John Sanko, "Colorado Only State Ever to Turn Down Olympics," *Denver Rocky Mountain News*, October 12, 1999.
4. Melious, et al, "Planning for the Olympics," [2–3]; Olympic Committee, *Final Report*, 15; Fred Tupper, "Moscow, Lake Placid Awarded '80 Olympics," *New York Times*, October 24, 1974; Howard Riley, "Remembering the Bid for 1980," in *25 Years After: Celebrating a Silver Anniversary, Lake Placid News and Adirondack Daily Enterprise*, February 11, 2005, 4; George Christian Ortloff and Steven C. Ortloff, *Lake Placid: The Olympic Years, 1932–1980: A Portrait of America's Premier Winter Resort* (Lake Placid, NY: Macromedia, 1976), 187–189; Carroll, *Winter Olympics*, 13–16.

5. Tupper, "Moscow, Lake Placid Awarded '80 Olympics"; Hearing before the Subcommittee on International Organizations and Movements of the Committee on Foreign Affairs, House of Representatives, 93rd Congress, 2nd Session on H. Con. Res. 443 and H. Con. Res. 444, May 8, 1974 (Washington: U.S. Government Printing Office, 1974), 3, 4, 17, 18, 29; Lake Placid Olympic Organizing Committee, "Information on the 1980 Olympic Winter Games Prepared by the Lake Placid Olympic Organizing Committee for the Congress of the United States," August 1975, 5, 6–7, 49–52; Dave Anderson, "Lake Placid's Olympic Perspective," *New York Times*, August 7, 1976; "Sierra Club Vies," *Lake Placid News*, November 23, 1976; Carroll, *Winter Olympics*, 38–40.
6. The Organizing Committee for the XI Winter Games, *The XI Winter Olympic Games: Sapporo, 1972* (1973), 120–132, 303–308; Fell, interviewed by Miller, 1–2; Paul Montgomery, "Lake Placid Already Aglow with 1980 Olympics Fever," *New York Times*, October 30, 1974; Carroll, *Winter Olympics*, 17.
7. Lake Placid Olympic Organizing Committee, "Information on the 1980 Olympic Winter Games by the Lake Placid Olympic Organizing Committee for the Congress of the United States," Exhibit 'B,' August 1975, v, viii, ix, x. See also Hearing before the Subcommittee on International Organizations and Movements of the Committee on Foreign Affairs, House of Representatives, 93rd Congress, 2nd Session on H. Con. Res. 443 and H. Con. Res. 444, May 8, 1974 (Washington: U.S. Government Printing Office, 1974), 4; and Bernard Adler, interviewed by Catherine Woodruff for the XIII Winter Games 1980 Oral History, November 29, 1979, [5], in MSS 90 Oral History, folder 4.
8. "Information on the 1980 Olympic Winter Games," 5; Carroll, *Winter Olympics*, 17–18.
9. Hearing before the Subcommittee on International Organizations and Movements, 21; Elizabeth Rider and Nancy Pepin, chap. 1: "Lake Placid: An Overview," in *Lake Placid and the 1980 Olympic Games*, [2–4]. On the oft-stated significance of the tourism and recreation industry, see William Colby, interviewed by Elizabeth Perry for the XIII Winter Games 1980 Oral History, January 18, 1980, [1–2], in MSS 90 Oral History, folder 1; Phil Laroque, interviewed by Miller for the XIII Winter Games 1980 Oral History, January 21, 1980, [8], in MSS 90 Oral History, Folder 1; O'Neill, interviewed by Perry, [14]; James Brooks, interviewed by Perry for the XIII Winter Games 1980 Oral History, January 18, 1980, 2, 26, in MSS 90 Oral history, folder 4; and Lake Placid Mayor Robert Peacock, interviewed by Derrick Holt for the XIII Winter Games 1980 Oral History, April 6, 1979, 3, in MSS 90 Oral History, folder 6. The exception to the rule of recreational development was the burgeoning prison industry in the Adirondacks. The Olympics, in fact, facilitated prison-building, as the Olympic Village in Ray Brook would be converted into a federal minimum-security prison. See XXXI Olympic Winter Games Committee, *Final Report*, 155–156.

10. John F. Maxwell, Raymond J. Richardson, and Patricia S. Olin, *The Economic Impact on the Lake Placid Area of Hosting the 1980 Winter Olympic Games* (Plattsburgh, NY: Technical Assistance Center, State University of New York, 1974), 1, 7, 17, 27–29. For a later discussion of the expected economic impact of the Olympics, including coverage of such issues as inflation and real estate values, see William R. Saunders and Richard E. Prusko in conjunction with the Technical Assistance Center, *XIII Olympic Winter Games, Lake Placid, 1980: Economic Impact Study* (August 1977). During the games, the New York Department of Labor expected a peak workforce of three thousand people. See Thomas O'Neill, interviewed by Perry for the XIII Winter Games 1980 Oral History, November 1, 1979, [5], in MSS 90 Oral History, folder 3.

11. Anderson, "Lake Placid's Olympic Perspective"; XIII Olympic Winter Games Committee, *Final Report*, 24–27, 32, 38–39, 45, 47–48, 63–66; Hearing before the Subcommittee on International Organizations and Movements, 22. See also Melious et al., "Planning for the Olympics," [7–8].

12. Melious et al., "Planning for the Olympics," [3, 4, 8]; Carol O'Connor, "Attitudes and Perceptions of Lake Placid Residents toward Hosting the 1980 Winter Olympic Games," chap. 3 in *Lake Placid and the 1980 Olympic Games*, 4, 6, 8, 10. See also "Area Backs Games."

13. Melious et al., "Planning for the Olympics," in *Lake Placid and the 1980 Olympic Games*, [1, 3, 7, 8]; O'Connor, "Attitudes and Perceptions of Lake Placid Residents," in *Lake Placid and the 1980 Olympic Games*, [2–3]. Wilkins was later fired from the Organizing Committee for alleged nepotism, and later prosecuted for fraud. On Wilkins, see Montgomery, "Lake Placid Already Aglow with 1980 Olympic Fever"; Glenn Fowler, "Already, Questions in Lake Placid," *New York Times*, March 13, 1977; John Kifner, "At Lake Placid, Optimism," *New York Times*, February 12, 1979; and Jane Gross, "Lake Placid: A Community Disrupted by Olympic Fever," *New York Times*, August 12, 1979. For more on residents' skepticism about the economic benefits of the Olympics, see Black Brook resident Margaret Forrester's letter to the editor, *Adirondack Daily Enterprise*, February 17, 1977, in which she warned, "Remember, someone has to pay for progress"; Matt Clark, interviewed by Linda Frame for the XIII Winter Games 1980 Oral History, April 24, 1979, 3, 4, 8, in MSS 90 Oral History, folder 4; and Shirley Seney, interviewed by Holt for the XIII Winter Games 1980 Oral History, April 26, 1979, 2, 3, 6, 11, 16, 21, in MSS 90 Oral History, folder 7.

14. Melious et al., "Planning for the Olympics," [7]; O'Connor, "Attitudes and Perceptions of Lake Placid Residents," [6, 3, 7].

15. O'Connor, "Attitudes and Perceptions of Lake Placid Residents," [7]; Harold Faber, "Getting There May Not Be Half the Fun for Fans at Lake Placid Olympics," *New York Times*, June 25, 1978; Jeff Platsky, "Olympic Pre-Games Tie up L. P. Traffic," *Adirondack Daily Enterprise*, February 12, 1979; Barbara Basler,

"Lake Placid on Eve of Olympics: 'Let the Games Begin,'" *New York Times*, February 11, 1980.

16. Melious et al., "Planning for the Olympics," [3, 5]; O'Connor, "Attitudes and Perceptions of Lake Placid Residents," [3, 7, 9]; Edna Broquist, interviewed by Perry for the XIII Winter Games 1980 Oral History, December 12, 1979, 15, in MSS 90, folder 4, Shirley Senney, interviewed by Holt for the XIII Winter Games 1980 Oral History, April 26, 1979, 21, in MSS 90 Oral History, folder 7. St. Lawrence University's poll got responses from many more men than women, mainly because the pollsters sought opinions from heads of households. In the smaller sample size of women respondents, women were found to be more ambivalent and resistant to the Olympics. The report's authors attributed that disparity to men's greater appreciation for sports and greater involvement in government, business, and sporting organizations. See O'Connor, "Attitudes and Perceptions of Lake Placid Residents," [2]. The desire to keep Lake Placid small, isolated, and attractive was widely held, and not only among women. Jim Rogers, owner of local radio station WBIZ, said, "I hope it is a beautiful town always." See Rogers, interviewed by Holt for the XIII Winter Games 1980 Oral History, April 17, 1979, [10], in MSS 90 Oral History, folder 3. Matt Clark, town clerk of North Elba, believed that the Olympics had outgrown Lake Placid: "I thought we were too small of a community to handle the Olympics. The Olympics are big business now." See Clark, interviewed by Frame, 2.

17. Hearing before the Subcommittee on International Organizations and Movements, 7, 10.

18. "After All, It's His Baby," *Lake Placid News*, November 2, 1978; Dave Anderson, "Olympic Update from Lake Placid," *New York Times*, February 9, 1978; XIII Olympic Winter Games Committee, *Final Report*, 38–39.

19. Rider and Pepin, "Lake Placid: An Overview," [1, 7–8]; Melious et al., "Planning for the Olympics," [6]; LPOOC, "Information on the 1980 Olympic Winter Games," 54–55; Maxwell et al., *The Economic Impact on the Lake Placid Area*, 1; Saunders and Prusko, *XIII Olympic Winter Games, Lake Placid, 1980: Economic Impact Study*, 56; James Kevlin, "APA Approval in Concept Asked for Olympics Jump," *Plattsburgh Press-Republican*, October 18, 1976; "Public hearing upon the construction of 90- and 70-meter ski jump facilities," November 19, 1976, in Adirondack Park Agency Records, series 3: MS 80-2, box 17: Olympic Hearings—Proceedings for 70 and 90 meter ski jump; "Environment vs. Development" (editorial), *Lake Placid News*, January 13, 1977. On proponents' economic argument, see also William Colby, interviewed by Elizabeth Perry for the XIII Winter Games 1980 Oral History, January 18, 1980, [3], in MSS 90 Oral History, folder 1; Donald Wheeler, interviewed by Elizabeth Perry for the XIII Winter Games 1980 Oral History, October 26, 1979, [5], in MSS Oral History, folder 3; James Brooks, interviewed by Perry for the XIII Winter Games 1980 Oral History, January 18, 1980, 2, 26, in MSS Oral History, folder 4; and

Tom Long, interviewed by Miller for the XIII Winter Games 1980 Oral History, n.d., 9, in MSS Oral History, folder 5.

20. "Balloon to Mark Ski Jump Height," *Lake Placid News*, May 20, 1976; Ray Lopez, interviewed by William C. Brayden for the XIII Winter Games 1980 Oral History, November 15, 1979, [9], in MSS Oral history folder 1. On appreciation of the ski jumps' beauty, see also Jay Rand to the editor, *Lake Placid News*, June 3, 1976; "Public hearing upon the construction of 90- and 70-meter ski jump facilities," November 19, 1976, 85, in Adirondack Park Agency Records, series 3: MS 80-2, box 17: Olympic Hearings—Proceedings for 70 and 90 meter ski jump; Gun Kaas Christiansen of Oslo, Norway, to the editor, *Lake Placid News*, June 3, 1976; Woody Cole, "Woody Cole Views the 90-Meter Ski Jump," *Lake Placid News*, October 7, 1976; Ray Lopez, interviewed by William C. Brayden for the XIII Winter Games 1980 Oral History, November 15, 1979, [9], in MSS 90 Oral History, folder 1; Paul Goldenberger, "Soaring Towers Relieve a Spartan Tradition," *New York Times*, February 19, 1980; and "Ski Jumps Are great Architectural and Tourist Triumph" (editorial), *Adirondack Daily Enterprise*, October 15, 1980.

21. LPOOC, "Application for Project Permit for 90 & 70 Meter Ski Jump Facility; Including Submission of Material for Conceptual Plan Review; Project Proponents: Lake Placid Winter Olympics, Inc., Town of North Elba Parks and Playground District," October 1976, III-1, 3, 5, 7, IV-2, 16, 32, in Adirondack Park Agency Records, series 3: MS 80-2, box 17: Olympic Hearings—Proceedings for 70 and 90 meter ski jump; "Public hearing upon the construction of 90- and 70-meter ski jump facilities," December 7, 1976, n.p., in Adirondack Park Agency Records, series 3: MS 80-2, box 17: Olympic Hearings—Proceedings for 70 and 90 meter ski jump; Sasaki Associates, Inc., *Final Environmental Impact Statement for 1980 Olympic Winter Games* (Washington, D.C.: Economic Development Administration, Department of Commerce, 1977), 171–173, 180; "Tower on Trial," *Lake Placid News*, October 20, 1976.

22. LPOOC, "Application for Project Permit for 90 & 70 Meter Ski Jump Facility," III-3, 21; Adirondack Park Agency, "Adirondack Park Land Use and Development Plan and Recommendations for Implementation," March 6, 1973, 2–3; Sasaki Associates, *Final Environmental Impact Statement*, 175–176; Margaret Dewey to the editor, *Lake Placid News*, January 20, 1977.

23. Sasaki Associates, *Final Environmental Impact Statement*, 188–189; LPOOC, "Application for Project Permit for 90 & 70 Meter Ski Jump Facility," III-12, 20–22, in Adirondack Park Agency Records, series 3: MS 80-2, box 17: Olympic Hearings—Proceedings for 70 and 90 meter ski jump.

24. Sasaki Associates, *Final Environmental Impact Statement*, 189, 192, 195; Harold Faber, "John Brown's Grave and Olympics Clash," *New York Times*, November 1, 1976.

25. "Public hearing upon the construction of 90- and 70-meter ski jump facilities," December 21, 1976, 1640–1642, 1657, in Adirondack Park Agency Records, series 3: MS 80-2, box 17: Olympic Hearings—Proceedings for 70 and 90 meter ski jump; "Jay Jump Pops Up," *Lake Placid News*, December 29, 1976.
26. Emmett N. O'Brien, "Intervale Jump Site Holds Key to Olympics," *Lake Placid News*, January 5, 1977. As evidence that the ski-jump debate made strange bedfellows, Robert J. Kafin had also represented environmental groups in the case against Ton-Da-Lay, while here he represented the LPOOC, which came under fire from those same groups.
27. "Balloon to Mark Ski Jump Height."
28. Emmett B. O'Brien, "Big Labor, Government Press Sierra Club," *Lake Placid News*, December 1, 1976.
29. "Olympic Plans of Lake Placid Are Criticized," *New York Times*, November 11, 1976; O'Brien, "Big Labor, Government Press Sierra Club"; "Sierra Club Vies"; Melious et al., "Planning for the Olympics," [6]. For more on Dumont, see Melious et al, "Planning for the Olympics," [5–7].
30. Gary Spence, "ADK Pulls Out," *Lake Placid News*, November 23, 1976; "Sierra Club's 'Standing' Questioned," *Lake Placid News*, November 24, 1976; O'Brien, "Big Labor, Government Press Sierra Club;" Melious et al., "Planning for the Olympics," [6]; J. Vernon Lamb, Jr., interviewed by Elizabeth Morrison for the XIII Winter Games 1980 Oral History, April 9, 1979, [2], in MSS 90 Oral History, folder 5; Schwartz, "The Olympics and the Environment," 24.
31. Melious et al, "Planning for the Olympics," [7]; Robert Allen, interviewed by Brayden for the XIII Winter Games 1980 Oral History, November 1, 1979, [10], in MSS 90 Oral History, folder 1.
32. "Public hearing upon the construction of 90- and 70-meter ski jump facilities," November 19, 1976, 53, in Adirondack Park Agency Records, series 3: MS 80-2, box 17: Olympic Hearings—Proceedings for 70 and 90 meter ski jump; Sasaki Associates, *Final Environmental Impact Statement*, 666–669; "Public hearing upon the construction of 90- and 70-meter ski jump facilities," November 29, 1976, n.p.; Mary E. Barrie, "Jay Jump?" *Lake Placid News*, July 8, 1976; "Alternate Site Aired in Albany for Placid Olympic Ski Jump," *Plattsburgh Press-Republican*, December 28, 1976; "Ski Jump Hearings Closed; Decision Expected Jan. 10," *Adirondack Daily Enterprise*, December 29, 1976; "Jay Jump Pops Up"; Steve Adamek, "Site Proposal No Threat to Olympics: Berle," *Plattsburgh Press-Republican*, December 31, 1976; Emmett N. O'Brien, "Intervale Jump Site Holds Key to Olympics," *Lake Placid News*, January 5, 1977.
33. Melious et al., "Planning for the Olympics," [6]; "Ski Jump Site Impact Aired"; "Jump-Site Foes Use Caution," *Lake Placid News*, January 20, 1977; "Public hearing upon the construction of 90- and 70-meter ski jump facilities," December 27, 1976, 1785–1786, 1796, in Adirondack Park Agency Records,

series 3: MS 80–2, box 17: Olympic Hearings—Proceedings for 70 and 90 meter ski jump.

34. State of New York Executive Department, Adirondack Park Agency, "Decision in the Matter of the application of Lake Placid 1980 Winter Olympic Games, Inc., and Town of North Elba as Trustees for Public Parks and Playgrounds District of the Town of North Elba, for Conceptual Approval Pursuant to Section 809 of the APA Act and 9 NYCRR 581.5 (f)," [January 1977], 1, 6–10; Charles Decker, "Intervales Is Approved for Games; Suits to Block Jump Are Threatened," *Adirondack Daily Enterprise*, January 10, 1977; Steve Adamek, "APA Decision Today Crucial to Olympics," *Plattsburgh Press-Republican*, January 10, 1977; "Council to Appeal Decision," *Adirondack Daily Enterprise*, January 11, 1977; Decker, "APA Commissioners Differed on Jump Vote but Are Undivided on Conclusion," *Adirondack Daily Enterprise*, January 11, 1977.

35. Melious et al., "Planning for the Olympics," [7]; "APA Vote Victory for Olympics" (editorial), *Plattsburgh Press-Republican*, January 13, 1977; Richard Nordwind, "Behind Jump-Site Decision Lies a Vale of Fears," *Lake Placid News*, January 13, 1977; Gary Spencer, "Legal Action Discussed," *Lake Placid News*, January 13, 1977; Susan Doolittle, "APA Rejected Report's Terms," *Adirondack Daily Enterprise*, January 14, 1977; Evelyn Outcalt, "Olympic Plans Hit by 2 Groups," *Adirondack Daily Enterprise*, January 17, 1977; Norbert Walker, "Carey Angered by Berle 'No' Vote on Ski Jump OK," *Plattsburgh Press-Republican*, January 19, 1977; O'Brien, "Governor Gets Tough to Assure 1980 Games," *Lake Placid News*, January 26, 1977.

36. Robert Flacke, interviewed by Morrison for the XIII Winter Games 1980 Oral History, April 19, 1979, [1–2, 4–8, 10, 12, 13]. See also Richard Estes, interviewed by Morrison for the XIII Winter Games Oral History, April 16, 1979, [1, 4–7]; and Persico, interviewed by Morrison, 3–5, 8–9.

37. Robert Flacke, interviewed by Morrison, [1–2, 4–8, 10, 12, 13]. See also Richard Estes, interviewed by Morrison, [1, 4–7]; and Persico, interviewed by Morrison, 3–5, 8–9.

38. "Ad'k Council Delays Suit," *Adirondack Daily Enterprise*, January 19, 1977; "Jump-site Foes Use Caution"; "Sierra Club, State Agree—APA Says No," *Lake Placid News*, April 21, 1977; Steve Adamek, "Park Agency Approves 2 Olympics Ski Jumps," *Plattsburgh Press-Republican*, April 23, 1977; "Intervale—the Jumping-off Place," *Lake Placid News*, April 26, 1977; "80's Olympic Ski Jumps Receive Final Approval," *New York Times*, April 26, 1977.

39. "Olympic Update," *Lake Placid News*, November 3, 1977; "Contractor Said Bankrupt," *Adirondack Daily Enterprise*, May 25, 1978; Allen, interviewed by Brayden, [10]; "Brown Farm" (editorial), *Lake Placid News*, June 1, 1978; Faber, "Construction Progressing for '80 Olympics," *New York Times*, June 4, 1978; "40-Ton Steel Truss Installed at Ski Jump," *Plattsburgh Press-Republican*,

October 13, 1978; Mary Feiss, "The Money Games," *Adirondack Life* XI (January/February 1980): 27; Michael Strauss, "New 70-Meter Jump Opens in Lake Placid," *New York Times*, December 31, 1978; Jeff Platsky, "Olympic Pre-Games Tie up L. P. Traffic," *Adirondack Daily Enterprise*, February 12, 1979.

40. Wheeler, interviewed by Perry, [5]; Goldenberger, "Soaring Towers Relieve a Spartan Tradition"; Laura Viscome, "Making the Most of Olympic Venues," *Lake Placid News*, August 21, 1980; "Panel Praises New York Tourism Promotions," *Adirondack Daily* Enterprise, October 2, 1980; "Ski Jumps Are Great Architectural and Tourist Triumph"; "Jumps Score Profit," *Adirondack Daily Enterprise*, November 20, 1980; Lena Williams, "For Lake Placid, Torch Still Glows," *New York Times*, February 27, 1981; "Checking out the Olympic Ski Jumps," *Lake Placid News*, December 19, 1985.

41. Richard Persico, interviewed by Morrison for the XIII Winter Games 1980 Oral History, April 19, 1979, 10, in MSS 90 Oral History, folder 6; Vera E. Littlejohn to the editor, *Adirondack Daily Enterprise*, October 15, 1980; Warren Kane, interviewed by Miller for the XIII Winter Games 1980 Oral History, January 29, 1980, 22 (italics added), in MSS 90 Oral History, folder 5.

42. XIII Winter Olympic Games Committee, *Final Report*, 124–133.

43. Glenn Fowler, "Already, Questions in Lake Placid," *New York Times*, March 13, 1977; David Bird, "1980 Olympic Work Roiling Lake Placid," *New York Times*, September 12, 1977; Broquist interviewed by Perry, 3, 13; Neil Chaffie, "Dollars, Fame, and the Games," *Lake Placid News*, 9 November 1978; Peacock, interviewed by Holt, 8–9. On economic impact, see also John Kifner, "At Lake Placid, Optimism," *New York Times*, February 12, 1979.

44. See Orloff, "LPOOC Has Map Problems," *Adirondack Daily Enterprise*, September 3, 1976; Lamb interviewed by Morrison, [1]; "From Bid to Ground Breaking," *Lake Placid News*, April 26, 1977; Nordwind, "Sierra Club's Trek to Trouble," *Lake Placid News*, August 31, 1977; Jeff Kelly, "Sierra Club Files Suit to Halt Road Work Near Cascade Lakes," *Plattsburgh Press-Republican*, February 25, 1978; Harold Faber, "Dispute over a Road in Adirondacks Heads to Court," *New York Times*, March 12, 1978; Faber, "Lake Placid Road Job Resuming," *New York Times*, July 3, 1978; Stephen J. Roberts, "Coalition Says Whiteface Trails Illegal," *Lake Placid News*, July 12, 1979; Jeff Platsky, "Ad'k Council Gets Go-Ahead," *Adirondack Daily Enterprise*, October 3, 1979; "Whiteface Mt. Suit Plans Proceeding," *Adirondack Daily Enterprise*, December 4, 1979; Roberts, "DEC Sued for Widening Whiteface Trails," *Lake Placid News*, December 6, 1979; Roberts, "State Is Sued over Widening Trails on Whiteface," *Lake Placid News*, May 29, 1980; Roberts, "State to Yield on Whiteface Trails," *Lake Placid News*, June 5, 1980; Ed Stransenback, "Adirondack Council Content with Widened Trails," *Plattsburgh Press- Republican*, March 5, 1980; Schwartz, "The Olympics and the Environment," 25–26, 71–72.

45. Steven R. Weisman, "US Is Seeking to End Outlays in Olympic Aid," *New York Times*, April 14, 1979; Harold Faber, "Olympic Site Faults Charged," *New York Times*, April 25, 1979; "Olympics Cost Put at 6 Times Estimate," *New York Times*, June 17, 1979; Red Smith, "Beautiful Dreamers," *New York Times*, December 9, 1979; Barbara Basler, "Quality of Olympics Work Questioned," *New York Times*, January 1, 1980; "Contractor Files Suit against LPOOC," *Lake Placid News*, January 3, 1980; Anderson, "A Village Welcomes the World and All of Its Problems," *New York Times*, February 10, 1980; "US Refuses Funds to Avert Olympic Unit's Bankruptcy," *New York Times*, May 17, 1980; Harold Faber, "$4.3 Million Deficit in Winter Olympics," *New York Times*, October 5, 1980; Senate Acts to Help Reduce Winter Olympics Debt," *New York Times*, December 12, 1980; XIII Winter Olympic Games Committee, *Final Report*, 145, 218, 219–222; Ned P. Rauch, "The Boys of Winter," in *Adirondack Life* XXXVI (January/February 2005): 16.
46. James M. Odato, "Owners Condemn Hotel Rates," *Adirondack Daily Enterprise*, May 25, 1978; William Oscar Johnson, "The Olympic Getaway," *Sports Illustrated*, April 9, 1979, 23; Jane Gross, "Lake Placid: A Community Disrupted by Olympic Fever," *New York Times*, August 12, 1979; Basler, "Lake Placid Braces Itself as Olympic Games Near," *New York Times*, November 19, 1979; Basler, "Lake Placid on Eve of Olympics: 'Let the Games Begin'"; William Doolittle, "Outsiders Blamed for Price Gouging," *Adirondack Daily Enterprise*, February 19, 1980.
47. Basler, "No Business Boom in Lake Placid Area," *New York Times*, February 19, 1980; William Doolittle, "Outsiders Blamed for Price Gouging."
48. Feiss, "The Money Games," 28.
49. Lou Reuter, "Lake Placid: The Last Games of Their Kind"; Lee Manchester, "Through the Shadow of the Cold War, Lake Placid's Winter Olympics Went on"; and Brittany Bombard, "Platsky: It Was the Best Time I Had in My Life," in *25 Years After*, 9, 10, 21–22.
50. "Wilmington Fears Health Hazard from Whiteface Garbage," *Lake Placid News*, February 13, 1980. See also Stephen J. Roberts, "Regional Landfill System Would Do away with Local Dumps," *Lake Placid News*, April 10, 1980; Charles Assetta to the editor, *Lake Placid News*, February 11, 1982; "Whiteface to Pay Fair Share at W'ton Dump," *Lake Placid News*, February 25, 1982; and Erin Doolittle, "Is Recycling Answer to Trash Problem?" *Lake Placid News*, December 9, 1987.
51. Basler, "Buses for Olympics Run into Snags," *New York Times*, February 12, 1980; Jeff Platsky, "Problems with Shuttle Bus System Reduced," *Adirondack Daily Enterprise*, February 19, 1980; Red Smith, "Lake Placid: The Deserted Village," *New York Times*, February 25, 1980; XIII Winter Olympic Games Committee, *Final Report*, 148–155; Lindall Lumsden, "For Bus Foul-ups, Afteruse

Planning State Faults LPOOC," *Lake Placid News*, April 17, 1980; "Lake Placid's Fiasco Revisited" (editorial), *New York Times*, May 20, 1980.

CONCLUSION

1. "Building Boom in Adirondack Park Tests Laws to Preserve Wilderness," *New York Times*, August 4, 1987; Elizabeth Kolbert, "Will the Adirondacks Secede to Vermont?" *New York Times*, November 23, 1990; Sarah Lyall, "Region and State at Odds over Adirondack Park," *New York Times*, October 26, 1989; Sam Howe Verhovek, "Governor Offers Plan to Preserve Wilderness Site," *New York Times*, October 11, 1991; "Adirondacks in Peril" (editorial), *New York Times*, June 16, 1992.
2. "Building Boom in Adirondack Park Tests Laws to Preserve Wilderness," *New York Times*, August 4, 1987; Sam Howe Verhovek, "For 100 Years, 'Forever Wild' and Forever in Dispute," *New York Times*, May 19, 1992.
3. Elizabeth Kolbert, "Residents Resent Huge Park's Plans," *New York Times*, December 26, 1986; Kolbert, "Will the Adirondacks Secede to Vermont?" *New York Times*, November 23, 1990.
4. Kolbert, "Panel Will Examine Land Use in Adirondack Park," *New York Times*, January 29, 1989; Lyall, "Region and State at Odds over Adirondack Park"; Lyall, "Adirondack Park Panel Seeks Balanced Future," *New York Times*, November 25, 1989; Verhovek, "Effort to Insure Adirondacks' Future Leads Planner to Albany and Beyond," *New York Times*, January 8, 1989; Verhovek, "Panel Urges State Purchases to Save Adirondack Wilderness," *New York Times*, May 5, 1990; Verhovek, "Cuomo Accused of Abandoning the Adirondacks," *New York Times*, September 24, 1990; Verhovek, "Cuomo Plans to Scale Back Proposal for Adirondacks," *New York Times*, August 14, 1991; Verhovek, "Governor Offers Plan to Preserve Wilderness Site"; Verhovek, "For 100 Years, 'Forever Wild' and Forever in Dispute"; "Adirondacks in Peril" (editorial). For more, see Barbara McMartin, *Perspectives on the Adirondacks: A Thirty-Year Struggle by People Protecting Their Treasure* (Syracuse: Syracuse University Press, 2002), parts 2 and 3; and Philip G. Terrie, *Contested Terrain: A New History of Nature and People in the Adirondacks*, 2nd ed. (Syracuse: Syracuse University Press, 2008), chap. 9.
5. Katherine Q. Seelye, "Rising Temperatures Threaten Fundamental Change for Ski Slopes," *New York Times*, December 12, 2012; Auden Schendler and Lukas Haynes, "Ski Mountains Move to Stop Climate Change," *High Country News*, September 4, 2013, accessed January 3, 2017, http://www.hcn.org/wotr/ski-mountains-take-an-important-step-to-stop-climate-change; Lauren Glendenning, "Climate Change and Ski Resorts, Part 1: Industry Wants Congress to Do More," *Summit Daily*, December 21, 2013; Glendenning, "Climate Change and Ski

Resorts, Part 2: Industry Looks to Take Lead on Issue," December 31, 2013; Glendenning, "Climate Change and Ski Resorts, Part 3: The 'Terrified' Versus the 'Sanguine,'" January 2, 2014; Lynch, "As Climate Changes, Poor Winters Hurt Adirondack Tourism," *Adirondack Almanack*, February 2, 2016, accessed January 9, 2017, http://www.adirondackalmanack.com/2016/02/white-stuff-green-stuff.html; Lynch, "The Winter That Wasn't," *Adirondack Explorer*, May 10, 2016, accessed January 9, 2017, http://www.adirondackexplorer.org/stories/dealing-climate-change.

6. Dean Baker quoted in Chris Knight, "Winter Takes a Vacation, Impacting Businesses," *Adirondack Daily Enterprise*, December 31, 2016.

7. Jenkins quoted in Lisa W. Foderaro, "Savoring Bogs and Moose, Fearing They'll Vanish as the Adirondacks Warm," *New York Times*, December 1, 2011. For more on climate change in the Adirondacks and New York State, see Dina Cappiello, "Scientists Study Climate Change in Adirondacks," *Baltimore Sun*, September 8, 2002, accessed January 7, 2017, http://articles.baltimoresun.com/2002-09-08/news/0209080038_1_high-peaks-adirondacks-warming; Jenkins, *Climate Change in the Adirondacks: The Path to Sustainability* (Ithaca: Comstock Publishing Associates, 2010); New York State Energy Research and Development Authority, *Responding to Climate Change in New York State: Technical Report*, November 2011, and *Climate Change in New York State: Updating the 2011 ClimAID Risk Information Supplement to NYSERDA Report 11–18 (Responding to Climage Change in New York State)*, September 2014; Chris Knight, "Experts: Adirondacks Should Prepare for Climate Change," North Country Public Radio, November 23, 2012, accessed January 9, 2017, http://www.northcountrypublicradio.org/news/story/20944/20121123/experts-adirondacks-should-prepare-for-climate-change; Lynch, "Climate Change Threatens Adirondack Boreal Species," *Adirondack Almanack*, September 15, 2015, accessed January 9, 2017, http://www.adirondackalmanack.com/2015/09/climate-change-threatens-adirodnack-boreal-species.html; Lynch, "Adirondack Climate Change: Deluges in the Forecast," *Adirondack Almanack*, April 24, 2016, accessed January 8, 2017, http://www.adirondackalmanack.com/2016/04/adirondack-climate-change-deluges-forecast.html; Tom Woodman, "Climate Change Will Affect the Adirondacks in Various Ways" (column), *Adirondack Daily Enterprise*, April 28, 2016; Lynch, "Dealing with Climate Change," *Adirondack Explorer*, May 10, 2016, accessed January 9, 2017, http://www.adirondackexplorer.org/stories/dealing-climate-change.

INDEX

Adirondack Civic League, 65
Adirondack Club and Resort, 1–2, 204
Adirondack Council, 162, 180, 188–93, 197
Adirondack Daily Enterprise, 146, 167–68
Adirondack Defense League, 163, 249n31
Adirondack League Club, 14–15
Adirondack Life, 198–99
Adirondack Mountain Authority, 95–99, 101, 232n18, 235n42, 235n44
Adirondack Mountain Club, 77, 100, 111, 122, 162, 188–90
Adirondack Park: campgrounds in, 41, 47, 50–52, 59; creation of, 15–16, 18; electrification of, 50; map of (1940), 9; map of (1980), 103; modern facilities in, 37–38; as modern wilderness playground, 1–10, 59; natural phenomena and visitors to, 57–60; popularity of, 54, 68, 74; resorts in, 17; usefulness of, 48, 110, 121, 123, 186–87; Whiteface Mountain, view from, 62–63, 66; Whiteface Mountain Memorial Highway construction, 64–74; Whiteface Ski Center and, 80. *See also* Forest Preserve; forever wild concept
Adirondack Park, exurbanization of. *See* second-home communities, planned development of

Adirondack Park Agency (APA): Adirondack Club and Resort and, 1; Casiers and, 163–65, 244n53; compared to Nazi state, 152–53, 248–49n28; creation of, 128–34; criticism of, 162, 179, 204, 254n58; exurbanization and, 137; Horizon Corporation and, 139–40, 142–44, 160–62, 167–69, 245–46n7; land categories designated by, 150–51, 156, 160–61, 187–88, 190–91; Loon Lake Estates and, 255n60; Olympic Winter Games (1980) and, 169–72, 175, 181–84, 186, 189–93, 195–97; residential development and, 203–4; threatened, 253–54n56; Ton-Da-Lay case and, 144–45, 149–53, 156, 159, 160–69; Wambat Realty case and, 254–55n60; zoning plans and, 107, 128–29, 133–34
Adirondack Park Agency v. Ton-Da-Lay, 165–67
Adirondack Park Association, 109, 129
Adirondack Park Local Government Review Board, 152, 162, 254n58
Adirondack Park Private Land Use and Management Plan, 133, 152–53, 156, 159, 167
Adirondack Park State Land Master Plan, 132
Adirondack Rail Company, 14
Adirondack Solidarity Alliance, 203

268 INDEX

Adirondack Wilderness Camp, 123
Adirondack Wilderness Committee, 78
Algonquin Peak, 188–89
Allen, Robert, 189–90
Altamont, New York, 144–47, 161–62, 168, 171, 186
Amell, Roger, 1
amendment 4 (skiing), 77–79, 81
amendment 7 (1927), 63–68, 91
American Legion, 64
American Scenic and Historic Preservation Society, 20
APA. *See* Adirondack Park Agency
APA Act, 131, 161, 166–67, 181, 254n60
Appalachian Mountain Club, 78
Arizona, 143
Armour, 146
Association for the Protection of the Adirondacks: APA Act and, 131; bobsled run controversy and, 24, 27–30, 35, 190; creation of, 19; forever wild concept and, 24, 48, 53; replaced by middle-class movement, 106, 116; support for amendment 4, 78–79; support for APA against Ton-Da-Lay, 162; well-heeled and genteel reputation of, 53, 72, 120, 137; Whiteface Mountain development and, 67–68, 72, 81
Association for the Protection of the Adirondacks v. Alexander MacDonald, Conservation Commissioner, 28–29, 36, 52–53
Atmospheric Sciences Research Center, 92
Ausable River, 91, 98, 185, 193
Au Sable Valley, 71

back-to-the-woods movement, 117–19
Bald Mountain, 88
Barnes, Jonathan F., 163
Bassett Mountain, 190–91
Battle, Gordon, 20–21
Battle Ground Camp, 44–45
Beamish, Richard, 249n28
Belleayre Mountain Ski Center, 86, 99, 101

Bentley, William L., 147, 149
Beyer, Libby, 149
Biesemeyer, Peter T., 249n28
Big Moose, 75
Blue Mountain Lake, 17
Blue Mountain Range, 82
Botkin, Daniel, 5
Boy Scouts, 148
Brandon, New York, 18–19
Brenna, Dick, 95–96
Brereton, Henry E.H., 52–54
Broadway comparison, 70–72, 88, 94, 101, 195–96
Broquist, Edna, 180
Buck, Harold, 140
Burdet, William, 23
Bureau of Forest Recreation, 114
Bureau of Land Management, 153
Burton, Chester L., 127
Buyce, Gerald, 99

Caldwell, New York, 44
California, 119
campgrounds, development of, 36–60; centralization and, 43; destructive effects of, 110, 122–24, 242n38, 242n40, 243n44; domestication, 43–44, 49–50, 59, 114, 136, 222n14; electric lights installed, 49–50; on Forest Preserve lands, 40–43, 94; forever wild provision and, 36–39, 243n44; New York State and, 36–38, 59, 82; popularity of camping and, 48–49; public health concerns and, 41–47, 55–57; rangers on motorcycles, 47–48; Temporary Study Commission on, 124–25; Whiteface development and, 69, 71
camping sprawl, 112–18, 124, 240n23
Canton, New York, 140
Carey, Hugh, 194, 200, 251n43
Carine, Frank V., 156–58
Casier, Audrey, 163–65, 249n31
Casier, Frank, 129–30, 163–65, 244n53
Catskills: Belleayre Mountain Ski Center in, 86, 99, 101; campgrounds

INDEX 269

in, 40, 48–49, 51, 113; closed due to fire risk, 57; Forest Reserve lands in, 15; ski trails in, 75; visitors to, 107–8
CBS. *See* Columbia Broadcasting System
CCC. *See* Civilian Conservation Corps
Chamonix, France, 22, 175–76
Champlain Valley, 110–11
Chester, New York, 45
Citizens' Northway Committee, 110–11, 117
Citizens to Save the Adirondack Park, 135, 140–41
Civilian Conservation Corps (CCC), 54–55, 57–58, 71, 115
Clare, New York, 139
Clark, Claude J., 121
Clark, Matt, 259n16
climate change, 206–8
Climate Change in the Adirondacks (Jenkins), 207
Clinton County, New York, 129
Cobham, Geoffrey A., 167–68, 255n64
Colton, New York, 139–42
Columbia Broadcasting System (CBS), 32
Commercial Record, 154
Commission on the Adirondacks in the Twenty-first Century, 204
Connery Pond, 46
conservation: Adirondackers' moral ecology and, 7, 68, 126, 148–49, 155; Adirondack Park as experiment in, 35; recreational amendment and, 53–54; recreational development, balanced with, 5, 40, 48, 107; state's role in, 14–16, 19, 40, 116, 120; wise use of resources and, 40, 110, 221n7. *See also* conservationists
Conservation Commission (1911–1927), 15, 21, 38–43, 47
Conservation Department (1927–1970): Adirondack Mountain Authority's responsibilities transferred to, 101; bobsled run and, 29, 33; campgrounds, electrification of, 49–50; campsite construction, spruce saved for, 69; campsite creation along Whiteface highway, 71; campsite improvements and, 28, 113–15; CCC camps and, 54–55, 58; creation of, 48; fire danger and, 57; Hoffman amendment and, 235n42; Lake Placid, photo of work in, 26; *Manual for Caretakers*, 47; Northway and, 108; recreation as priority for, 21, 38, 48–52, 123; second-home owners' protests against, 56–57; ski centers, planning and administration of, 86; ski trail construction and, 75–77, 228n29; supply and demand problems, 107; survey (1960), 112–13; Whiteface Ski Center and, 79; World War II and, 58
conservationists: defined, 213n5; elitist reputation of, 70; environmentalists vs., 189–90; forest management and, 221n7; local business promotion and, 40; preservationists vs., 5; recreational development and, 65–66, 78–79. *See also* conservation
constitutional amendments, 7–8; amendment 4 (skiing), 77–79, 81; amendment 7 (1927), 63–68, 91; authorizing two new ski centers (1947), 85–86, 91; forever wild provision and, 2, 24; Hoffman amendment, 82–83, 99–100, 130, 235n42; recreational amendment, 52–54, 60
consumer culture, 39–40
Council of Conservationists, 109
Countryman, Stanley, 124
Crimes against Nature (Jacoby), 4
Cropper, Evelyn, 140
Crown Point, 46
Cuomo, Mario, 204

Davis, George D., 157
DEC. *See* Department of Environmental Conservation
deforestation, 38
deindustrialization, 3, 10, 23
Democratic Party, 53

democratization, 65–66, 73, 106–7, 109, 121
Denver, Colorado, 174
Department of Environmental Conservation (DEC): Conservation Department replaced by, 101; environmentalists/conservationists and, 193, 197; land-use restriction enforcement and, 168; Loon Lake Estates and, 255n60; mental appreciation of nature and, 191; Olympic preparations and, 197; ski-jump controversy and, 185; structure removal by, 187–88; Ton-Da-Lay and, 145–46, 148–50, 152–54, 157–60, 191, 247n19, 251n43
Department of Public Works, 32, 91, 108–9, 193, 197
Devlin, Art, 189
Dewey, Godfrey, 22–25, 27–31, 35, 190
Dewey, Margaret, 184
Dewey, Thomas E., 80, 87
Dexter, Orrando, 18
DeZalia, James, 102
Diamond, Henry L., 148, 156–59, 191
Dolbeck, Merrill, 151
Dominion Bridge, 193
Draper, Arthur, 93–95, 109–10, 237n8
Dreissigacker, John, 89
droughts, 38, 59
Dry Channel Pond, 147, 157–58
Dumont, James, 187–88, 192
Durant, William West, 14–15

Eagle Point Camp, 45
Economic Development Administration (EDA), 181–82, 197
economic growth: Olympics and, 34, 171–72, 176–77, 182, 192; park-making and, 3; public campsites and, 39; public vs. private recreational facilities, 97; recreational development and, 5, 58–59, 65–66, 82, 90–91; ski center development, promotion of, and, 100, 103–4; Ton-Da-Lay support and, 147

economy of North Country: campground development and, 38; private lands and, 136; tourism and, 3–4, 101–3, 121, 131, 149, 165, 176, 204–5; Whiteface Memorial Highway construction and, 63–64, 68; Whiteface ski centers and, 84, 88. *See also* unemployment in the Adirondacks
EDA. *See* Economic Development Administration
Elkins, Frank, 86
environmentalists: Adirondackers and, 126; APA and, 133, 170; conservationists vs., 189–90; destructive results of mass recreation and, 122–23; interests shared with Adirondackers, 207; loss of open space and, 117; Olympics and, 178, 180–82, 185–90, 192–93, 196–97; opposition to, 151, 178; private land development and, 137–38, 149, 168; proposal to convert Adirondack Park into national park and, 120
environmental justice, 207–8, 212n14
environmental movement, 6–7, 52, 117–19, 212n14
environmental politics in North Country, 4, 83, 104, 205. *See also* wilderness politics
Essex County, New York: APA legislation and, 129; campgrounds in, 43–46; deindustrialization and, 23; Hoffman amendment and, 99; Olympics and, 11, 24, 173, 176; tourism and, 121; unemployment in, 102; Whiteface Mountain Memorial Highway and, 61
Essex County Chamber of Commerce, 111
Essex County Republican, 131
Estes, Richard F., 156, 253–54n56
Esther Mountain, 85
extractive industries, 5, 8–10, 16, 27, 38, 40, 139. *See also* logging industry; mining industry

exurbanization of private lands. *See* second-home communities, planned development of

Fawn Ridge, 75
Federal Highway Act, 19
Federal Trade Commission, 143
Federation Internationale de Ski, 91
Fell, J. Bernard, 177
Felton, Carl, 146
Fine, New York, 153
fires, 38, 57, 59
Fish Creek Pond, 49, 52, 55–57, 114–15, 132
Flacke, Robert, 192–93
Flynn, Andy, 89
Foley, William J., 151
Follensby Pond, 56, 123
Forest, Fish, and Game Commission (1885–1911), 15
forest management, 221n7
Forest Preserve: Adirondack Park State Land Master Plan and, 132; bobsled run controversy and, 13; campgrounds, development of, 10, 40–43, 94–95, 105–6, 113, 125; Conservation Department and, 101; conservation of, 19, 24, 27, 29–30; creation of, 18; forever wild provision and, 7, 28, 35, 36, 53, 133, 168; highway construction through, 64, 72, 108–11, 133; Horizon Corporation and trespassing and tree cutting in, 143, 169; lands set aside as, 15–16; modern facilities in, 37, 39, 50, 55; natural phenomena and visitors to, 57–58; Olympic development and, 197; overcrowding in, 123–24; "real *wilderness* areas" in, 118; recreational development in, 40, 48, 50–52, 54, 57, 59, 81; ski trails in, 75–76, 100, 228n28; stronger oversight of, 119–120; Ton-Da-Lay plans and, 147, 158; vacationers, clout of, and, 115; World War II and, 58. *See also* Adirondack Park; campgrounds, development of; Catskills

forever wild concept: affirmation of, 52–53; APA Act and, 129, 131–33; Association for the Protection of the Adirondacks and, 19; bobsled run controversy and, 13, 27–29, 35; camping facilities and, 16, 36–39, 49, 54, 114–16; defined, 38–39, 132–33; mass recreation and, 40, 48, 57, 74–76, 79–80, 83, 121; modifications to, 60, 82, 98–99, 121–22, 132–33; New York State and, 204; Northway and, 108, 110–11, 133; Olympic development and, 197, 200; private land development and, 117, 135–37, 142, 168; second-home owners and, 117; ski-jump building and, 172; state constitution and, 2, 24; Whiteface Mountain Memorial Highway and, 63–64, 77–78; wilderness, meaning of, and, 201; wilderness uses and, 7; wildlife, disruption of, and, 85
Forsyth, Alfred S., 122
Fountain, William J., 100, 121, 235n42
Franklin County, New York: Adirondack Club and Resort and, 1; established resource uses and, 164–65; income, average weekly (1968), 169, 255n65; Loon Lake Estates development plans and, 255n60; population loss in, 169; recreational development, promotion of, and, 121; Rockefellers and, 18–19; Ton-Da-Lay and, 138, 143–46, 151, 154–55, 161–62, 164–65
Franklin County Chamber of Commerce, 148–49
Franklin Falls Trail, 75
Frenette, Jim, 171–72, 186
Friedman, Ruth and Ralph, 163

Gale, Brian, 163
Galehouse, Richard F., 185
Gallos, Phil, 147–48
gastroenteritis, 55–56
Gibbs, John T., 76
Gibson, James, 254n60

Gilbane Building Company, 183, 193
Gitlen, Philip G., 185
Glennon, Robert C., 163
Glens Falls, New York, 129
Gore Mountain, 75, 86, 91, 95–97
Gore Mountain Ski Club, 228n29
Gore Mountain Ski Resort, 99, 101, 228n29, 234n36
Grady, Noreen, 94–95
Graves, Guy A., 162–63
Great Camps, 14–15, 18, 28, 78, 108
Great Depression, 30–31, 50–51, 53–54, 57, 65
Greene, Frederick Stuart, 68–70, 72, 190
Grover, Richard, 142
Gruppe, Karl, 123, 242n40
G.S. Franklin & Co., 40

Harrietstown, New York, 162
Harriman, Averell, 90–92
Harris, Glenn, 129, 143, 152
Harroff, Richard A., 155
Harvey, David, 5
Hawaii, 119
Hayes, Jeremiah M., 165–66
Hays, Samuel, 213n5
Hearthstone Point Camp, 44–45, 50–52, 54, 132
Hepburn, Lionel P., 142
Heritage Village, 154–55
Herkimer County, 88
Hetch Hetchy controversy (1913), 5
high modernism, 5, 37, 133
High Peaks: Olympic Winter Games (1932) and, 11, 23, 25; Olympic Winter Games (1980) and, 172, 179, 182–91; ski site development and, 91; Whiteface Mountain Memorial Highway and, 61
Hochschild, Harold K., 120, 127
Hoff, Philip, 154–55
Hoffman amendment, 82–83, 99–101, 130, 235n42
Hoffman Mountain, 82–83, 99–101, 235n42

Horizon Corporation, 135–36, 138–44, 160–62, 167–69, 245–46n7
Howard, William G., 42
Hullar, Theodore, 180

industrialization, 15
Innsbruck, Austria, 174
International Olympic Committee (IOC), 22–25, 174–75
Intervale, New York, 22, 34, 171–72, 176, 200
Intervale ski-jump controversy, 172, 181–96, 261n26
IOC. See International Olympic Committee
Iron Mountain, 156
Izaak Walton League, 66, 123

Jacoby, Karl, 4, 7
Jay, New York, 190
Jenkins, Jerry, 207
Jensen, Jerome W., 158, 191
Jerome, Jerome V., 65–66
Jerry, Harold A., 127–28
John Brown's Farm, 185
Johnson, Lyndon B., 119
Joint Legislative Committee on Environmental Management, 143
Joint Legislative Committee on Winter Tourists, 90–91
Jones, R. Courtney, 180
Jordan River, 156, 160–61
Jordan River Estates, 160–61

Kafin, Robert J., 186, 261n26
Kane, Warren, 195, 196
Ketchledge, Edwin H., 188–89
Kiamesha Lake, 233n28
Kilborn, Thomas, 242n38
Kildare Club, 148, 162–63
Klein, Harold, 123
Kress, Peter E., 127

Lake Champlain, 46
Lake George, New York, 41, 44, 54, 101, 105, 123

INDEX 273

Lake George Association, 123
Lake George Battleground, 240n23
Lake Placid, New York: desire to keep small and attractive, 259n16; development surge in, 196–200; economy in, 102; Little Whiteface and, 92; Olympic Winter Games (1932) and, 11–13, 21–27, 31–35; Olympic Winter Games (1980) and, 171–89, 191, 195–96, 198–201; Whiteface Mountain Memorial Highway and, 61, 64, 75
Lake Placid Bid Committee, 174–76
Lake Placid Business Association, 180
Lake Placid Central School Board of Education, 180
Lake Placid Chamber of Commerce, 23, 87, 100
Lake Placid Club, 12–15, 22, 25, 30
Lake Placid Kiwanis Club, 22, 23
Lake Placid News, 65–66, 69, 80, 130–31, 147–48, 183, 199
Lake Placid Olympic Committee, 34
Lake Placid Olympic Organizing Committee (LPOOC): Adirondack Mountain Club and, 189; APA and, 192; challenges faced by, 200; Kafin and, 186, 261n26; new infrastructure and, 182–83, 186–87; Olympic Winter Games (1980), plans and preparations for, 175–78, 180, 196–98
Lake Placid Ski Club, 91
Lake Placid Ski Council, 76
Lake Tahoe, 119
Lamb, J. Vernon, Jr., 76–77, 183, 189
Lamora, Oliver, 18–19
Lamy, Margaret, 105–6, 121–22
Landolfe, Dave, 164–65
land-use policy: bobsled run controversy, 13, 24–25, 27–30, 35, 190; campgrounds, 117; continuing conflicts over, 201; mapping of lands and, 119; Northway construction conflict, 106–12, 117, 133; state vs. federal, 120; strict, champions of, 106; urban-rural links and, 3, 8; wealthy favored in, 18–19; zoning, 119, 126–27, 133–34.

See also Adirondack Park Agency; Intervale ski-jump controversy; second-home communities, planned development of
Lansing, John, 186
Lefkowitz, Louis, 161, 166
legibility: imposed on North Country, 6, 205; management of state parks and, 5–6, 37, 43, 46, 118; for recreation seekers, 37–38, 50, 59–60, 118; zoning and, 181
Lehman, Herbert, 57, 71
Lengyel, Henry W., 160
Lindsay, Robert, 243n44
Little, Frank, 139
Littlejohn, Vera E., 195
Little Whiteface Mountain, 83, 90–102
Little Whiteface Mountain Ski Center, 91–99; Adirondack Mountain Authority, criticism of, 96–97; chairlift on, 92–93, 232n18; construction of, 91–92; difficulty accommodating visitors to, 93–95; Hoffman amendment and, 101; snowfall on, inadequate, 95–96; snowmaking equipment installed on, 97–99, 233n28
logging industry, 18, 23, 27, 37, 38
Long Island, New York, 149, 247n22
Lookout Mountain, 86
Loon Lake Estates, Inc., 255n60
Lopez, Ramon, 183
Los Angeles, California, 118–19
LPOOC. *See* Lake Placid Olympic Organizing Committee

MacIntyre Range, 91
MacKenzie, Ronald, 182
Main, Robert, 90–91
Manning, Beverly, 198
Marble Mountain and Marble Mountain Ski Center, 79–80, 83–92, 100–101, 110
Martitsch, Karl, 182
mass recreation: consequences to land, 46–47, 57, 110, 122–23;

mass recreation *(continued)*
  Forest Preserve developed by New York State, 38; forever wild provision and, 24; during Great Depression, 51; during interwar years, 37; loss of wilderness and, 103; modern amenities and, 3; national trends toward, 19; in nineteenth century, 14; Olympics and, 23, 28–30, 34–35; organized park system and, 20; State Council of Parks and, 21
Maybell, Everett A., 131
McEwen, Robert, 176
McMorran, J. Burch, 237n8
McMullen, Sally, 82–83
McRobbie, Scott, 140, 142
mining industry, 18, 23, 37
Mirror Lake, 22
"Mission 66" (National Park Service), 112
Moberg, Per O., 124
modernism, 5–6. *See also* high modernism
modernization of campsites in Forest Preserve, 38–58
modern wilderness playground, Adirondack Park as, 1–10; Adirondack Club and Resort project, 1–2; consequences of, 120; diversity and contested nature of park, 3–4; electrification and, 50; high modernism and, 5–6; North Country economy and, 3–4; obstacles to, 81; resource conservation and, 5; stakeholders in park's future and, 6–10, 59; state intervention and, 2, 3, 6, 8–10
Moon Hill Camp, 39
moral ecology: Adirondack Park creation and, 164; in *Crimes against Nature* (Jacoby), 4, 7–8; home rule and, 68, 126, 141, 143; Olympics, support for, and, 179; Paparazzo's appeal to, 148–49, 155
Morgan, J. Pierpont, 19
Morgenthau, Henry, Jr., 53
Morton, Lysle W., 123
Moses, Robert, 20, 41, 120

Mount Killington, Vermont, 96
Mount McKenzie, 91
Mount Whitney, 75
Mt. Van Hoevenberg, 11, 12, 32, 176, 195–197
Mt. Van Hoevenberg Bobsled Run, 29–30, 33, 75, 82, 95, 101
Mulholland, William G., 228n28
Murray, William H.H. "Adirondack," 16–17, 39
Murray's Fools, 16–17, 47, 113–14

Nash, Roderick, 7
Nassau County, New York, 247n22
National Broadcasting Company (NBC), 32
National Geographic report, 55
National Park Service, 112, 119
National Park Service Organic Act, 19
Nature Conservancy, 255n64
NBC. *See* National Broadcasting Company
Nelson, Sidney, 139, 165
Nevada, 119, 153
Newbold, Katherine H., 56–57, 123
Newcomb, New York, 129
New England, competition with, 76–77, 86, 95–96, 100
Newhouse, David L., 242n38
New Mexico, 143
New York City, 25, 53, 68
New York Conservation Association, 66
New York–New Jersey Trail Conference, 78
New York State: camping facilities and, 36–38, 222n14; conflicts with Adirondackers, 4, 118, 129–33, 137–38, 150, 153, 163; conservation role of, 14–16, 19, 40, 116, 122; forever wild concept and, 204; intervention in the Adirondacks, 2–3, 6, 8–10; legibility forced on North Country and, 5–6; Long Island study, 247n22; in lose-lose trap, 97; moral ecology of Adirondackers and, 7–8; Olympic funding from, 31; park management

and city planning, 8; recreational amendment, 52–54, 60; recreational development in Forest Preserve and, 38, 59; secession threats, 203; sued by Ton-Da-Lay, 253n55; urban-rural links in, 3, 8–10; Whiteface ski centers and criticism of, 84. *See also* recreation in New York State, re-creation of

New York State Association, 20

New York State Conservation Council, 140

New York Times, 34, 66–68, 70, 72–74, 86, 97, 102, 109–10, 131, 147, 150–51, 198, 203

Niagara Mohawk Power Corporation, 115

Nichols, Mrs. George, 155

Northampton Beach, 240n23

North Creek, New York, 75, 228n29

North Elba, New York: contaminated water in Connery Pond near, 46; Intervale ski jumps and, 22, 183–84, 195–96; Olympic development and, 198–200; Olympic Winter Games (1932) and, 11, 22–23, 27, 30–31, 33–34; Olympic Winter Games (1980), referendum on, and, 174–75

Northern Lumber Company, 139

North Hudson, New York, 43–44

North Notch, 75

Northway, 106–12, 117, 126, 133, 237n8

Oak Mountain, 75

O'Brien, Leo W., 120

Old Forge, New York, 75, 88

Old Forge–Malone line, 14

Oliver, Frank J., 122

Olympic Winter Games: competitive requirements for, 91; re-creation of recreation in New York State and, 13–22. *See also* Lake Placid, New York

Olympic Winter Games (1932), 11–13, 22–35; bid to IOC, 22–25; bobsled run, 11–13, 24–25, 27–30, 33–35, 197; ice arena building, 31–34; modernization of Lake Placid and, 12–13, 21–22, 25–27, 32–35; Olympic stadium building, 30–31, 33–34; skiing and, 75

Olympic Winter Games (1972), 175

Olympic Winter Games (1976), 174

Olympic Winter Games (1980), 169–202; APA and, 169–72, 175, 181–84, 186, 189–93, 195–97; economic growth goals and, 171–72, 176–77, 182, 192; economic impact of, 196, 198, 258n10; Lake Placid and, 171–89, 191, 195–96, 198–201; nostalgia for, 199; prison-building and, 257n9; problems caused by, 199–200; ski-jump controversy, 181–96, 261n26

O'Neill, Florence, 179

"Operation Outdoors" (U.S. Forest Service), 112

Operation Squeezo, 95, 97

Ordway, Samuel H., 53

Osborne, Lithgow, 74

overdevelopment in the Adirondacks, 105–34; APA, creation of, and, 129–33; campground sprawl, 112–18; Northway, 106–12; Temporary Study Commission and, 118–29

Owens Pond Camp, 45–46

Paine, Peter S., Jr., 95

Palisades Interstate Park Commission, 20

Palmer, Adam R., 165–66

Paparazzo, Louis, 144–50, 152–62, 165–69, 247n19, 248n28

parkway construction, 21. *See also* Northway; Whiteface Mountain Memorial Highway

Persico, Richard A., 161, 193

Pierce, Grace, 187

Pinchot, Gifford, 221n7

Plattsburgh, New York, 146

Plattsburgh Daily Press, 70–72, 74

Plattsburgh Press-Republican, 129, 149

Plattsburgh University, 176

Plumadore, Jan, Jr., 163

Plumb, Robert, 141

Porter, Fred L., 24, 52–54, 76, 79
Potter, Orlando B., 127
prison industry, 257n9
private land development. *See* residential development in Adirondack Park; second-home communities, planned development of
Prospect House, 17, 49
Prospect Mountain, 101
Prospect Mountain Veterans Memorial Highway, 101
Purdue, Richard B., 150–51, 254n58
Purdy, Robert R., 248–49n28

Quinlivan, James J., 55–56

railroad expansion, 14
Raquette Lake Railway, 14
Ray Brook, New York, 148, 163, 197, 257n9
Reagan, Ronald, 165
recreational amendment, 52–54, 60
recreational development in Adirondack Park: campgrounds, development of, 36–60; camping sprawl, 112–18; modern wilderness playground, 1–10; Olympic Winter Games (1932), 11–35; Olympic Winter Games (1980), 171–202; overdevelopment, 105–34; second-home communities, planned development of, 135–70; Whiteface Mountain, skiing on, 82–104; Whiteface Mountain transformation, 61–81
recreation in New York State, re-creation of, 13–22; American social and economic trends and, 13–14, 19–20; conservation efforts, 14–16, 19; Great Camps, 14–15; Murray's Fools, 16–17; park system organization, 20–22; resorts, 17; tensions with local residents, 17–19
residential development in Adirondack Park, 203–4. *See also* second-home communities, planned development of

resource management classification, 150–51, 156, 160–61, 190–91
Rochow, William and Janet, 131
Rockefeller, Laurance, 120
Rockefeller, Nelson, 95, 100, 105, 107, 120, 132, 152
Rockefeller, William A., 148
Rockefeller, William G., 14–15, 18–19, 108
Roden, William, 90, 131
Rogers, Jim, 259n16
Roosevelt, Franklin D., 24, 27, 31, 50, 53
Rothman, Hal, 154
route B vs. route C debate, 108–12
Rural Electrification Administration, 50

Sacandaga campsite, 43, 242n40
Sagamore Lodge, 15
Sagebrush Rebellion, 153
"Sally's Skiing," 82
Santa Clara, New York, 144
Santa Claus's Workshop, 87
Sapporo, Japan, 175–77
Saranac Lake, 25, 56, 64, 169, 233n28
Saranac Lakes Wild Forest, 148, 156
Sasaki Associates, 181, 184–85
"Save the Adirondack Park" (poster), 141
Schlickenrieder, Warren, 94
Schroon Lake, 33, 39, 45
Schroon-North Hudson Winter Sports Council, 100
Schroon River, 43–44
Scott, James, 5–6
Scott's Cobble, 75
second-home communities, planned development of, 135–70; APA and, 137, 139–40, 142–45, 149–53, 156, 159–69; Citizens to Save the Adirondack Park and, 135, 140; Horizon Corporation, 135–36, 139–44, 160–62; Ton-Da-Lay, 136, 138, 143–67; Vail, Colorado, and, 250n33; Wambat Realty case and, 254n60
second-home owners: APA Act and, 131–32; campsite development

opposed by, 56–57, 60, 81, 116–17; land-use policy and, 106; Northway debate and, 108; residential development plans and, 137, 168; zoning and, 126–28
Second National Conference on State Parks, 20
Seney, Shirley, 180, 198
Sentinel Range, 27
Shea, Jack, 198–99
Sierra Club: APA and, 162, 187–88; camping facilities and, 122–23; local sentiment toward, 178, 187; Northway debate and, 111, 117; Olympics and, 180, 189–93, 197; as Ton-Da-Lay opponent, 153–54
skiing: Little Whiteface Mountain Ski Center and, 90–102; Marble Mountain Ski Center and, 84–90; obstacles to, 82–84, 98–99, 104; transformation of Whiteface Mountain and, 74–81. *See also* Olympic Winter Games; Whiteface Mountain, skiing on
Smith, Alfred E., 21, 64
snowmaking equipment, 97–99, 233n28
snow trains, 75
Southbury, Connecticut, 154–55
Sparks, John M., 151
Speculator, New York, 75
spot zoning, 166
Spring Pond Bog, 156
Spurney, Petr, 198
Stafford, Ronald, 100, 129
State Council of Parks, 21, 41, 107, 120
Steinhardt, Maxwell, 67–68
St. Lawrence County: APA opponents in, 153; Franklin County compared to, 169; Horizon's plans for second-home community in, 135, 139, 141–42, 160, 169; income, average weekly (1968), 169, 255n65; population increase in, 169, 255n65
St. Lawrence County Environmental Management Council, 140
St. Lawrence University, 177, 259n16
St. Moritz, Switzerland, 22, 31, 175–76

Stock, John, 127–28, 244n53
Strauss, Michael, 97
St. Regis Canoe Area, 148, 156–57
Students for Environmental Alternatives, 140
Sugarbush, Vermont, 96
Sugar Loaf, 75
Sullivan, Francis, 87

Tahoe Regional Planning Agency, 119
Taylor Pond, 61
Tea Party, 153
Temporary Study Commission on the Future of the Adirondacks, 118–29; APA and, 107, 128–29; campsite construction and, 114; campsite policy (new) recommended by, 124–25; national trend toward better land management and, 118–20; public hearings by, 105–6, 120–24; ski center development and, 125–26; zoning recommendation of, 126–29
Terrie, Philip, 39–40
Ticonderoga Fish and Game Club, 151
Ton-Da-Lay, 144–67; APA and, 144–45, 149–53, 156, 159, 160–69; Boston Tea Party comparison, 153, 248n23; DEC and, 145–46, 148–50, 152–54, 157–60, 191, 247n19, 251n43; Forest Preserve and, 147, 158; Franklin County, New York, and, 138, 143–46, 151, 154–55, 161–62, 164–65; Kafin and, 261n26; local support for, 138, 147; second-home communities, plans for, 136–37; state sued by, 253n55
Ton-Da-Lay v. Diamond, 161
Tongue Mountain Range, 54
Torrey, Raymond H., 72
tourism: acceptance of, 7–8; Adirondackers and, 17–18, 102–3, 118, 141–42, 164; deindustrialization and, 23, 37, 38, 40, 176; economic inequality and, 207–8; economy of North Country and, 3–4, 101–3, 121, 131, 149, 165, 176, 204–5; industry limitations, 10; infrastructure improvements and, 17;

tourism *(continued)*
  Olympics and opportunity for, 11–12, 23; promotion of, 39, 64–65, 77, 80, 126, 204; urbanization and, 14, 37
Tupper Lake, New York, 1–2, 75, 144, 146, 162, 204
Tupper Lake Chamber of Commerce, 144

unemployment in the Adirondacks, 88, 102, 121, 136, 146, 176, 231n10
Union Falls Overlook, 61
urbanization: forest destruction and, 15; state park planning and, 20–21, 58–59; tourism and, 14, 37
urban-rural links, 3, 8–10, 141, 149
U.S. Forest Service, 112, 119
U.S. Plywood-Champion Papers, 146
U.S. Route 9 Improvement Association, 111
Uzdzvinis, Joe, 121

Vail, Colorado, 250n33
Valley Run trail, 98
Valmont Village, 254n60
Vanderbilt, Alfred G., 15, 19, 108
Vanderwalker, David, 150, 153, 248n23
Van Norden, Ottomar H., 78–79
Vermont, 96, 100, 125, 154
Verner, Elliot K., 243n44

Walsh, M. I., 123
Wambat Realty, 254–55n60
Wargo, John, 191
War Production Board, 58
Warren County, New York, 41, 44–45, 54, 75, 129
Warrensburg Chamber of Commerce, 111
Washington County, New York, 129
Wells, Willis, 27, 68
White Brook, 85
Whiteface Act, 69
Whiteface Action Committee, 96–97
Whiteface Area Advisory Council, 79
Whiteface Area Ski Council, 77
Whiteface Authority Act, 80
Whiteface Castle, 61, 72–73
Whiteface Highway Commission, 80
Whiteface Mountain, and Olympic Winter Games (1980), 176, 197, 200
Whiteface Mountain, skiing on, 82–104; Little Whiteface, developing, 90–102; Marble Mountain Ski Center, 84–90; natural obstacles to, 82–84, 98–99, 104; New York State, criticism of, and, 84, 130
Whiteface Mountain, transformation of, 61–81; highway construction and opening, 64–70, 87; skiing on, promotion of, 74–81; summit of, 73; tunnel and elevator built into, 71–72; visitors to, 70–71
Whiteface Mountain Authority, 80, 85–86, 88, 91, 93–95
Whiteface Mountain Highway Commission, 69
Whiteface Mountain Memorial Highway, 10, 60–74; Adirondack Mountain Authority and, 95; Conservation Department and, 101; conservationists and, 190; construction of, 64–69; construction of, opposition to, 66–68, 70, 72, 81; construction of, passing of amendment for, 68; construction of, support for, 65–66, 72–73, 82; highway ending at summit, 73, 78; opening of, 69–70, 77, 87; recreational purpose of, 65; unattractive development along, 87–88; Whiteface Ski Center and, 80
Whiteface Mountain Ski Center, 79–80, 86–87, 95, 101
Whiteface Racing Trail, 76–77
Wikoff, Jack, 96
wilderness: definitions of, 108–9, 132, 201; domestication of, 37–38, 49, 73–74; loss of, 77–78, 90, 103, 110–11, 184, 188; mass recreation and, 68, 103; separation between city and, 67, 70–72, 79, 84, 123–24
Wilderness Act (1964), 119, 132, 158, 201
wilderness battles, history of, 2, 8–10. *See also* land-use policy

wilderness politics: campers' expectations and, 46; complex nature of, 138; continuing conflict caused by, 204; ownership of park and, 106; stakeholders in park's future and, 6–10. *See also* land-use policy
Wilderness Society, 111, 117, 187
wilderness zoning, 119
Wilkins, Jack, 178, 198
Wilmington, New York, 45–46, 61, 64, 80, 92, 199
Winter Olympics (1932 and 1980), 10
wise-use movement, 110

Wolfe, Douglas, 89
Wood, David E., 148–49
Woodshed Restaurant, 198
World Bobsled Championships (1961), 173
World University Games (1972), 173–74
World War I veterans memorial, 64
World War II, 34, 58, 75, 80

Zepa Consulting Co., 255n64
zoning, 119, 126–29, 133–34, 144–45, 162, 175, 181. *See also* spot zoning

JONATHAN D. ANZALONE was born in Glen Cove, New York. He has three degrees from the State University of New York system: a BA from Geneseo College, an MA from Binghamton University, and a PhD in history from Stony Brook University. He is currently the assistant director of the Center for News Literacy and a lecturer in the Stony Brook School of Journalism. He also teaches history courses for Stony Brook's School of Professional Development and Suffolk Community College's Social Sciences Department. He has been the recipient of the Presidential Fellowship from Stony Brook University and the Larry J. Hackman Research Residency from the New York State Archives Partnership Trust. He has contributed to the *Encyclopedia of American Environmental History*, *American Studies Journal*, and H-Net Reviews. He lives on Long Island with his wife, Sally, and his son, Benjamin.

www.ingramcontent.com/pod-product-compliance
Lightning Source LLC
Chambersburg PA
CBHW021654230426
43668CB00008B/622